AF832032

GeoJournal Library

Volume 133

Series Editor
Barney Warf, University of Kansas, Lawrence, KS, USA

Now accepted for Scopus! Content available on the Scopus site in summer 2021.

This book series serves as a broad platform for scientific contributions in the field of human geography and its sub-disciplines. The series, which published its first volume in 1984, explores theoretical approaches and new perspectives and developments in the field of human geography.

Some topics covered by the series are:

- Economic Geography
- Political Geography
- Cultural Geography
- Historical Geography
- Health and Medical Geography
- Environmental Geography and Sustainable Development
- Legal Geography and Policy
- Urban Geography
- Geospatial Techniques
- Urban Planning and Development
- Land Use Modelling
- and much more

Publishing a broad portfolio of peer-reviewed scientific books, GeoJournal Library invites book proposals for research monographs and edited volumes. The books can range from theoretical approaches to empirical studies and contain interdisciplinary approaches, case studies and best-practice assessments. The books in the series provide a great resource to academics, researchers and practitioners in the field.

Fivos Papadimitriou

Geo-Topology

Theory, Models and Applications

Springer

Fivos Papadimitriou
Mathematisch-Naturwissenschaftliche Fakultät,
Forschungsbereich Geographie
Universität Tübingen
Tübingen, Germany

ISSN 0924-5499 ISSN 2215-0072 (electronic)
GeoJournal Library
ISBN 978-3-031-48187-1 ISBN 978-3-031-48185-7 (eBook)
https://doi.org/10.1007/978-3-031-48185-7

© The Editor(s) (if applicable) and The Author(s), under exclusive license to Springer Nature Switzerland AG 2023, corrected publication 2024
This work is subject to copyright. All rights are solely and exclusively licensed by the Publisher, whether the whole or part of the material is concerned, specifically the rights of translation, reprinting, reuse of illustrations, recitation, broadcasting, reproduction on microfilms or in any other physical way, and transmission or information storage and retrieval, electronic adaptation, computer software, or by similar or dissimilar methodology now known or hereafter developed.
The use of general descriptive names, registered names, trademarks, service marks, etc. in this publication does not imply, even in the absence of a specific statement, that such names are exempt from the relevant protective laws and regulations and therefore free for general use.
The publisher, the authors, and the editors are safe to assume that the advice and information in this book are believed to be true and accurate at the date of publication. Neither the publisher nor the authors or the editors give a warranty, expressed or implied, with respect to the material contained herein or for any errors or omissions that may have been made. The publisher remains neutral with regard to jurisdictional claims in published maps and institutional affiliations.

This Springer imprint is published by the registered company Springer Nature Switzerland AG
The registered company address is: Gewerbestrasse 11, 6330 Cham, Switzerland

Paper in this product is recyclable

To the memory of my father Theodossios, who taught me how a place *can give* meaning *to life*

Preface

Both geography and topology are concerned with the study of spatial objects and patterns, albeit with different aims and by adopting different methods each. Broadly speaking, while topology is concerned with properties of spatial entities of any n-dimensional space, geography focuses on spatial entities, phenomena and processes that are encountered on the face of the earth.

It is more than likely that topologists are already aware of some of the concepts and methods presented here, but it is rather unlikely that they would have imagined the range of fascinating applications of these concepts and methods in geography.

Reversely, the relevance of topology to geography has hitherto remained largely unknown to geographers. And yet, neither geographers nor topologists have explored how and why topology might be significant for geography.

The domains of convergence of these two disciplines are identified in this book from the point of view of geography (i.e. in geographical analysis and modelling, geo-visualization, geographical education and philosophy of geography), revealing how concepts of topology (boundaries, networks, connectedness, knots, braids, Klein bottles, handles, Euler characteristics, Reeb graphs, fixed point theorems, homotopy, etc.) might have applications in landscape geography, biogeography, urban geography, social geography, GIS and geospatial technologies.

Part I introduces geo-topology with an account of how the "spatial turn" in geography steadily and imperceptibly cedes its place to a "topological turn" (Chap. 1).

Part II presents specific applications of topology in geography. As all GIS experts are already familiar with the word "topology", this part of the book begins with a review of the most widely known applications of topology in geo-informatics (Chap. 2), which is followed by characteristic applications of topology in geographical analysis, focusing on landscape boundaries (Chap. 3), networks of borders (Chap. 4), landscape complexity and resilience (Chap. 5).

Part III unveils the usefulness of topology for geo-visualization: from Euler numbers and Reeb graphs (Chap. 6) to dynamical systems (Chap. 7), knot theory and braid theory (Chap. 8). With these chapters, it is shown how hitherto disregarded but nevertheless central concepts of topology can be useful in geo-visualization.

Part IV suggests certain topological shapes and surfaces (e.g. Hopf links, the whirlpool function, *k*-connected spaces and cobordisms) as metaphors for the convergence of the geo-space with the cyber-space (Chap. 9), while also drawing the attention of geographers to the potential of ultrametric topologies for modelling social relationships in combined geo- and- cyber-spaces (Chap. 10).

Part V examines how topology might contribute to geographical analysis with applications in psychology (Chap. 11), geographical education and gaming (Chap. 12), and its repercussions for the epistemology (Chap. 13) and philosophy (Chap. 14) of geography. In the penultimate chapter in particular, it is explored how the topological perspective of spatial analysis differs from the geographical, whereas the last chapter exposes how topology permeates our everyday urban lifestyles and introduces the concept of "personal topology".

Possibly, alternative titles of this book might as well have been "topology for geography", "topological geography" or "geographical topology" and anyone of them would suit it well.

The ancient philosopher Iamblichus suggested in his "Protrepticus" to avoid avenues (that most people follow) and walk on paths instead ("Τὰς λεωφόρους ὁδοὺς ἐκκλίνων, διὰ τῶν ἀτραπῶν βάδιζε"). So the ultimate goal of writing this book is to open the path of topological thinking in geography.

This new path I named "Geo-Topology".

Whether for applications in geographical modeling, geo-visualization, geographical education or geo-philosophy, tapping into the enormous potential of topology may enrich and advance geographical theory and practice in unimaginably many ways. What is presented in this book is, hopefully, only the beginning.

Mathematisch-Naturwissenschaftliche Fakultät, Forschungsbereich Geographie
Universität Tübingen
Tübingen, Germany

Fivos Papadimitriou

Contents

The original version of this book was published with a spelling mistake in the author's last name. A correction to this book can be found at https://doi.org/10.1007/978-3-031-48185-7_15

Part I
Introduction

Chapter 1
The Topological Turn in Geography

Location (topos) is the greatest of all.
It contains everything.
“Μέγιστον τόπος. Άπαντα γάρ χωρεῖ”
(Thales, 640–546 b.C., in Diog. Laert. “Lives of Philosophers”)

1.1 From Geometry to the Topological Revolution

Research problems, theories and applications of topology remained confined within the disciplinary boundaries of mathematics for quite a long time. In fact, concepts and methods of topology were not used by scientists other than mathematicians and physicists, even well into the second half of the twentieth century. But the time of seclusion for topology corresponds to a remote era of the past, since we are nowadays immersed in a trove of topological concepts and applications and scientists from several scientific domains have become well-acquainted with topological ideas and methods. So much so, that topology bears manifold applications in domains that were previously unrelated to it: geography and geoinformatics, ecology, complexity, chemistry, biology and morphogenesis, electronics, computer science, telecommunications and many more.

It might appear as this expansion had been a rather unimportant issue to pay attention to. But it isn't: as a matter of fact, a great part of the technologies of our civilization are being created on topological premises, marking an unmistakeable shift from the “*geometric*” era (from the beginning of civilization up to 1980) to a “*topological*” era.

Geometry attracted the interest of humankind from the ancient Egyptians and the Greeks to the builders of the cathedrals in Europe and the cartographers and navigators of the past centuries. The emphasis that people laid on spatial measurements and metrics of the geographical space was driven by the need for activities such as precision in navigation, ship building, creation of military equipment, etc. In the “quantitative era”, geometry was as indispensable as algebra, much like statistics in the twentieth century. In the prior to the twentieth centuries, most scientific fields failed to implement topological concepts and methods. As a matter of fact, in many cases, topology served as an aid to gain deeper insights into geometry rather than as a

© The Author(s), under exclusive license to Springer Nature Switzerland AG 2023

F. Papadimitriou, *Geo-Topology*, GeoJournal Library 133,
https://doi.org/10.1007/978-3-031-48185-7_1

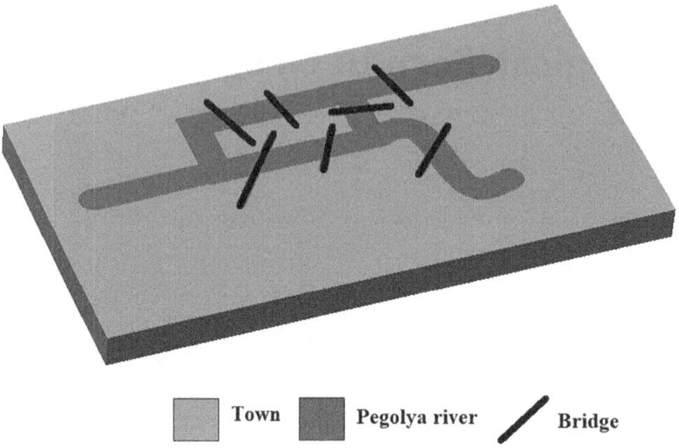

Fig. 1.1 A schematic representation of the Königsberg bridges problem: It is impossible to avoid crossing twice one of the bridges in a single journey that would cross them all

rigorous scientific domain of its own that would boldly demonstrate its own range of applicability to problems of everyday life.

Looking back at the non-topological periods of the history of mathematics, one may as well wonder how none of the great thinkers of the past ever thought of some topological concepts before Euler did when he was asked to solve the famous "Königsberg bridges" problem (Euler, 1752). This problem consisted in finding a way to cross all the bridges of the city of Königsberg (Fig. 1.1) by crossing each one of them only once and Euler proved in 1752 that this is impossible.

In the later half of the nineteenth century and with Poincaré establishing it as a discipline able to stand independently from the other fields of mathematics, topology entered an era of frenetic research that yielded remarkable results and gradually developed into a vivid branch of mathematics with its main field of study the qualitative properties of spaces and spatial entities.

Currently, and as concerns geography, topology is useful in the analysis of various geographical features such as borders, spatial networks, cellular maps, landscape structure, distributions of points, curves and polygons in geoinformatics etc. Its applicability was given a strong impetus within the theory and practice of GIS, to the extent that there is hardly any geographer nowadays that purports to carry out research in some domains of geography such as transport geography or quantitative landscape analysis without using concepts and methods of topology.

Probably, the foremost important examples of entanglement of topology with geography have been in efforts to solve geographical problems which turned into topological problems, such as the famous "Four Color Theorem" (proving that four colors suffice to color a planar map) and the "Travelling Salesman Problem" (TSP) on visiting places. Of these two geographical problems, the first was solved but the second remains still difficult to solve.

Solving the "Four Colors Problem" with a legendary proof that required some 10,000 diagrams and 1000 h of computer time, Appel and Haken proved in 1976 that four colors suffice to assign different colors to the regions of a 2D map (Appel & Haken, 1977a, b, c). Some cases however are excluded, i.e. regions joining at one point, one-boundary countries and maps broken in two or more pieces and, as a result, the "Four Color Theorem" is restricted to cases where there are exactly three boundary lines at meeting points.

The TSP has obvious repercussions for route optimisation in geographical analysis and logistics (Grötschel & Holland, 1991; Gu et al., 2015; Xia et al., 2016; Santos & Oliveira, 2021; Carvalho et al., 2022) and consists in finding the shortest possible route that connects n cities (which have different distances among them), so that each city is visited only once and the trip ends up at the city from which it begun. To get a glimpse of the high complexity involved in this problem, if n cities had to be visited, then there would be $[(n\text{-}1)!]/2$ possible routes: for $n = 4$ cities there are only 3 routes, for $n = 5$ there are 12, for $n = 10$ there are 181,440 possible routes and for $n = 30$ there are as many as 4,420,880,996,869,850,977,271,808,000,000 that is close to $4{,}4 \times 10^{30}$. With such an explosive growth (Cook, 2014), the computational complexity class of this problem is "NP-hard" (non-determnistic polynomial hard). Hence, creating algorithms for the solution of the TSP remains particularly challenging (Christofides & Eilon, 1972; Lenstra & Kan, 1975; Kirkpatrick & Toulouse, 1985; Laporte & Martello, 1990; Laporte et al., 1996; Dorigo & Gambardella, 1997; Brezina Jr & Čičková, 2011; Kumbharana & Pandey, 2013). Strangely perhaps, empirical research has discovered that humans can solve it unexpectedly quickly and accurately (Gärling, 1989; MacGregor & Ormerod, 1996; Dry et al., 2012).

Although topology adopts primarily qualitative approaches to examine spatial objects and their relationships, it is also endowed with quantitative methods for analysing those qualitative properties. The concept of dimension is probably the most remarkable qualitative characteristic of both geographical and topological objects and its quantification revealed some surprising results.

The dimension of a physical or mathematical entity is its foremost important characteristic and the study of dimensions is a par excellence field of topology. The discovery that certain mathematical objects might have non-integer dimensions (Mandelbrot, 1977, 1982) was one of the major discoveries of twentieth century topology. This led to the theory of fractals which, precisely as their name suggests, are shapes with non-integer (fractional) dimensions. For instance, the fractal dimension of the "Koch curve" is 1.2619 (in between 1 that is the dimension of a line and 2 that is the dimensions of the plane). Fractals present one of the most remarkable junctions of topology and geography nowadays. It is worth noticing that the very first observation of a fractal in geography (and, indeed, as considered by many, the trigger to the intensive research in fractals that followed) was the cartographic discovery (Richardson, 1961) that while the Portuguese measured their border with Spain as 987 km, the Spanish measured it as much longer: 1214 km. Shortly after that, the "coastline paradox" was revealed with the seminal paper title "How long is the coast of Britain? Statistical self-similarity and fractional dimension" (Mandelbrot, 1967). The paradox consists in the phenomenon by which the measured length of a coastline increases with the spatial resolution of measurements (i.e. the length of the coastline of Great Britain measured at a resolution of 100x100km is 2800 km long, but it is 3400 km if measured at a resolution of 50x50km). It is precisely the expression "fractional dimension" in Mandelbrot's paper that reveals the purely topological context of the paper (the non-topological part deals with self-similarity). The concept of fractals was initially used to analyse shapes and dimensions of urban settlements and geographical boundaries (Longley & Batty, 1989a, b; Longley et al., 1991).

With these works opened up a whole new field of research in geography. To date, fractals have been used in geographical analysis repeatedly, i.e. to analyse street networks (Rodin & Rodina, 2000; Buhl et al., 2006; Cardillo et al., 2006; Murcio et al., 2015; Jiang & Brandt, 2016; Bocewicz et al., 2016; Lu et al., 2016; Jiang, 2019; Zhang et al., 2022; Sui et al., 2022) and to simulate morphologies of urban expansion (Batty & Longley, 1987; Frankhauser, 1998a, b; De Keersmaecker et al., 2003; Cooper, 2005; Chen, 2013; Chen & Wang, 2013; Klinkhamer et al., 2017; Boeing, 2018; Purevtseren et al., 2018; Chen & Huang, 2019; Ma et al., 2020; Malishevsky, 2022). Interestingly, the findings of such researches are not always confined to theoretical explanations. If, for instance, a city expands linearly along distinct radii emanating from its central district, then the fractal dimension of that city is higher compared to another of which the shape is relatively round. But this type of expansion means it is very likely the buildings that are located along those linear expansions have easier access to green areas and/or empty spaces and thus "fractality" (even up to a certain extent only) can explain city function. For instance, the fractality of cities that are more concentrated around a historical centre (i.e. as many European cities are) tends to be lower compared with that of cities that expand in sprawls, i.e. American cities (Batty & Longley, 1987; Frankhauser, 1998a, b).

Growing from strength to strength, and with their value being recognized by increasingly more scientists worldwide, applications of topology seem to expand by leaps and bounds, not only crossing the borders of other scientific domains, but also into the public sphere and in everyday life. This is nothing short of a *topological revolution* which, aside of science, can also be perceived by considering the topological terms that entered the vocabulary of billions of people: "links", "connected", "network", "borders" etc.

1.2 The "Topological Turn"

The milder form of this revolution appeared in the late 1980s, with a "topological turn" that followed the "spatial turn" in geography and other sciences (Shields, 2012). This "turn" entailed an emphasis on concepts of nonlinear dynamical systems, such as attractors and bifurcations, catastrophe theory and abrupt changes in chaotic dynamics paired with an emphasis on flows and fluidity (Deleuze & Guattari, 1987; De Landa, 2002; Massumi, 2002; Mackenzie, 2005; Kantor, 2005; Rotman,

2012; Lury et al., 2012), as "the idiom of the topological offers a language for articulating the instabilities and fluctuations of state territory" (Harvey, 2012, p.77).

In sociology, one of the typical examples of the topological turn can be identified in Bourdieu's theory of social space (Guy, 2018), as well as in the context of Latour's actor-network theory (Latour, 1987). In the "era of connection" (Weinberger, 2003), we talk about "nomadic", "decentred", "situated" subjectivities (Massumi, 1992), as "Homo fractalis" emerges from the virtual world alongside with a "diaspora of networks" that include attractors, fractals and intelligence (Baudrillard, 2005, p.36,59).

In philosophy, the topological context of Heidegger's philosophy was highlighted by Malpas (2006, 2012) and the "becoming topological of culture" has been brought forth as a social-cultural process by Lury et al. (2012), since topological surfaces constitute an essential characteristic of our time, bearing the marks of smoothness, fluidity and adaptability.

Notions of topology also appeared in the context of psychology with Lacan's theories (Lacan, 2006; Phillips, 2013) and with interpretations of the processes of "deterritorialization" and "reterritorialization" of populations and individuals (Deleuze & Guattari, 1987).

Oddly however, despite geography and topology being both disciplines that explore spatial entities, despite the significance of concepts of topology in geoinformatics and GIS theory and practice, and despite its central position in any effort to decipher the complexity of anything spatial (Papadimitriou, 2020a, b, c), topology showed up rather late in human and cultural geography (Paasi, 2011; Allen, 2011; Secor, 2013a; Martin & Secor, 2014).

The relationships between human geography and topology can be examined from within different perspectives. For instance, topology has been associated with diverse issues and fields of interest to human geography such as culture (Shields, 2012, 2013), regions with intense historical significance (Giaccaria & Minca, 2011), flows and fluidity in spatial processes (Mol & Law, 1994; Shields, 1997; Häkli, 2008; Harker, 2014) and urban experience (Secor, 2013). Furthermore, topology has been a source of inspiration for wider human geographic explanations. While, for instance, acknowledging that some topological surfaces can be useful in examining relational structures (Malpas, 2012; Secor, 2013), it has been proposed (Belcher et al., 2008, p.503) "to seize the potential of emergence, the potential of topological transformation, to undermine the apparent fixity of current geometries of power" and, possibly representing one of the most interesting developments in the theory of cultural studies, the term "cultural topology" was coined by Lury (2013).

However, it would not be far from the truth to consider that the "topological turn" has been a never fully-accomplished and never widely-accepted sideway of the history of science in the twentieth century. But the magnitude and intensity of the topological turn of the previous century is dwarfed by the explosive growth of networks in the twenty-first century which was also aided by the massive expansion of information and communication technologies.

References

Allen, J. (2011). Making space for topology. *Dialogues in Human Geography, 1*, 316–318.

Appel, K., & Haken, W. (1977a). Every planar map is four colourable. I. Discharging. *Illinois Journal of Mathematics, 21*(3), 429–490.

Appel, K., & Haken, W. (1977b). Every planar map is four colourable. II. Reducibility. *Illinois Journal of Mathematics, 21*(3), 491–567.

Appel, K., & Haken, W. (1977c). Solution of the four color map problem. *Scientific American, 237*(4), 108–121.

Batty, M., & Longley, P. A. (1987). Fractal-based description of urban form. *Environment and planning B: Planning and Design, 14*(2), 123–134.

Baudrillard, J. (2005). *The intelligence of evil or the lucidity pact*. Berg.

Belcher, O., Martin, L., Secor, A., Simon, S., & Wilson, T. (2008). Everywhere and nowhere: The exception and the topological challenge to geography. *Antipode, 40*(4), 499–503.

Bocewicz, G., Jardzioch, A., & Banaszak, Z. (2016). Modelling and performance evaluation of fractal topology streets network. In *Distributed computing and artificial intelligence, 13th international conference* (pp. 483–494). Springer International Publishing.

Boeing, G. (2018). Measuring the complexity of urban form and design. *Urban Design International, 23*(4), 281–292.

Brezina, I., Jr., & Čičková, Z. (2011). Solving the travelling salesman problem using the ant colony optimization. *Management Information Systems, 6*(4), 10–14.

Buhl, J., Gautrais, J., Reeves, N., Solé, R. V., Valverde, S., Kuntz, P., & Theraulaz, G. (2006). Topological patterns in street networks of self-organized urban settlements. *The European Physical Journal B-Condensed Matter and Complex Systems, 49*, 513–522.

Cardillo, A., Scellato, S., Latora, V., & Porta, S. (2006). Structural properties of planar graphs of urban street patterns. *Physical Review E, 73*(6), 066107.

Carvalho, H. S., Pilastri, A., Novais, R., & Cortez, P. (2022). RanCoord—A random geographic coordinates generator for transport and logistics research and development activities. *Software Impacts, 14*, 100428.

Chen, Y. (2013). A set of formulae on fractal dimension relations and its application to urban form. *Chaos, Solitons & Fractals, 54*, 150–158.

Chen, Y., & Huang, L. (2019). Modeling growth curve of fractal dimension of urban form of Beijing. *Physica A: Statistical Mechanics and its Applications, 523*, 1038–1056.

Chen, Y., & Wang, J. (2013). Multifractal characterization of urban form and growth: The case of Beijing. *Environment and Planning B: Planning and Design, 40*(5), 884–904.

Christofides, N., & Eilon, S. (1972). Algorithms for large-scale travelling salesman problems. *Journal of the Operational Research Society, 23*(4), 511–518.

Cook, W. J. (2014). *The travelling salesman problem. Mathematics at the limits of computation*. Princeton University Press.

Cooper, J. (2005). Assessing urban character: The use of fractal analysis of street edges. *Urban Morphology, 9*(2), 95.

De Keersmaecker, M. L., Frankhauser, P., & Thomas, I. (2003). Using fractal dimensions for characterizing intra-urban diversity: The example of Brussels. *Geographical Analysis, 35*(4), 310–328.

De Landa, M. (2002). *Intensive science and virtual philosophy*. Continuum.

Deleuze, G., & Guattari, F. (1987). *A thousand plateaus: Capitalism and schizophrenia*. University of Minnesota Press.

Dorigo, M., & Gambardella, L. M. (1997). Ant colonies for the travelling salesman problem. *Biosystems, 43*(2), 73–81.

Dry, M., Preiss, K., & Wagemans, J. (2012). Clustering, randomness and regularity: Spatial distributions and human performance on the traveling salesperson problem and minimum spanning tree problem. *The Journal of Problem Solving, 4*(1), 1–17.

Euler, L. (1752 [1741]). Solutio problematis ad geometriam situs pertinentis, 1736. *Commentarii Academiae Scientiarum Imperialis Petropolitanae, 8*, 128–140.

Frankhauser, P. (1998a). Fractal geometry of urban patterns and their morphogenesis. *Discrete Dynamics in Nature and Society, 2*(2), 127–145.

Frankhauser, P. (1998b). The fractal approach. A new tool for the spatial analysis of urban agglomerations. *Population: An English selection, 10*(1), 205–240.

Gärling, T. (1989). The role of cognitive maps in spatial decisions. *Journal of Environmental Psychology, 9*(4), 269–278.

Giaccaria, P., & Minca, C. (2011). Topographies/topologies of the camp: Auschwitz as a spatial threshold. *Political Geography, 30*, 3–12.

Grötschel, M., & Holland, O. (1991). Solution of large-scale symmetric travelling salesman problems. *Mathematical Programming, 51*(1–3), 141–202.

Gu, W., Liu, Y., Wei, L., & Dong, B. (2015). A hybrid optimization algorithm for travelling salesman problem based on geographical information system for logistics distribution. In *LISS 2014: Proceedings of 4th international conference on logistics, informatics and service science* (pp. 1641–1646). Springer.

Guy, J.-S. (2018). Bourdieu in hyperspace: From social topology to the space of flows. *International Review of Sociology, 28*, 510–523.

Häkli, J. (2008). Regions, networks and fluidity in the Finnish nation-state. *National Identities, 10*, 5–22.

Harker, C. (2014). The only way is up? Ordinary topologies of Ramallah. *International Journal of Urban and Regional Research, 38*, 318–335.

Harvey, P. (2012). The topological quality of infrastructural relation: An ethnographic approach. *Theory, Culture & Society, 29*(4/5), 76–92.

Jiang, B. (2019). A topological representation for taking cities as a coherent whole. In *The Mathematics of Urban Morphology* (pp. 335–352). Springer.

Jiang, B., & Brandt, S. A. (2016). A fractal perspective on scale in geography. *ISPRS International Journal of Geo-Information, 5*(6), 95.

Kantor, J.-M. (2005). A tale of bridges: Topology and architecture. *Nexus Network Journal, 7*(2), 13–21.

Kirkpatrick, S., & Toulouse, G. (1985). Configuration space analysis of travelling salesman problems. *Journal de Physique, 46*(8), 1277–1292.

Klinkhamer, C., Krueger, E., Zhan, X., Blumensaat, F., Ukkusuri, S., & Rao, P. S. C. (2017). Functionally fractal urban networks: Geospatial co-location and homogeneity of infrastructure. *arXiv*, 1712.03883.

Kumbharana, S. N., & Pandey, G. M. (2013). Solving travelling salesman problem using firefly algorithm. *International Journal for Research in Science & Advanced Technologies, 2*(2), 53–57.

Lacan, J. (2006). *Ecrits*. (B. Fink, Trans.). Norton.

Laporte, G., & Martello, S. (1990). The selective travelling salesman problem. *Discrete Applied Mathematics, 26*(2–3), 193–207.

Laporte, G., Asef-Vaziri, A., & Sriskandarajah, C. (1996). Some applications of the generalized travelling salesman problem. *Journal of the Operational Research Society, 47*, 1461–1467.

Latour, B. (1987). *Science in action*. Harvard University Press.

Lenstra, J. K., & Kan, A. R. (1975). Some simple applications of the travelling salesman problem. *Journal of the Operational Research Society, 26*(4), 717–733.

Longley, P. A., & Batty, M. (1989b). Fractal measurement and line generalization. *Computer & Geosciences, 15*(2), 167–183.

Longley, P. A., Batty, M., & Shepherd, J. (1991). The size, shape and dimension of urban settlements. *Transactions of the Institute of British Geographers (New Series), 16*(1), 75–94.

Longley, P. A., & Batty, M. (1989a). On the fractal measurement of geographical boundaries. *Geographical Analysis, 21*(1), 47–67.

Lu, Z., Zhang, H., Southworth, F., & Crittenden, J. (2016). Fractal dimensions of metropolitan area road networks and the impacts on the urban built environment. *Ecological Indicators, 70*, 285–296.

Lury, C. (2013). Topological sense-making: Walking the Mobius strip from cultural topology to topological culture. *Space and Culture, 16*(2), 128–132.

Lury, C., Parisi, L., & Terranova, T. (2012). Introduction: The becoming topological of culture. *Theory, Culture & Society, 29*(4–5), 3–35.

Ma, D., Guo, R., Zheng, Y., Zhao, Z., He, F., & Zhu, W. (2020). Understanding Chinese urban form: The universal fractal pattern of street networks over 298 cities. *ISPRS International Journal of Geo-Information, 9*(4), 192.

MacGregor, J. N., & Ormerod, T. (1996). Human performance on the traveling salesman problem. *Perception & Psychophysics, 58*, 527–539.

Mackenzie, A. (2005). The problem of the attractor: A singular generality between sciences and social theory. *Theory, Culture & Society, 22*(5), 45–65.

Malishevsky, A. (2022). Fractal analysis and its applications in urban environment. In *System analysis & intelligent computing: Theory and applications* (pp. 355–376). Springer International Publishing.

Malpas, J. (2006). *Heidegger's topology: Being, place, world*. MIT Press.

Malpas, J. (2012). Putting space in place: Philosophical topography and relational geography. *Environment and Planning D: Society and Space, 30*(2), 226–242.

Mandelbrot, B. (1967). How long is the coast of Britain? Statistical self-similarity and fractional dimension. *Science, 156*(3775), 636–638.

Mandelbrot, B. (1977). *Fractals: Form, chance, and dimension*. W. H. Freeman and Company.

Mandelbrot, B. (1982). *The fractal geometry of nature*. W. H. Freeman and Company.

Martin, L., & Secor, A. J. (2014). Towards a post-mathematical topology. *Progress in Human Geography, 38*(3), 420–438.

Massumi, B. (1992). *A user's guide to capitalism and schizophrenia: Deviations from Deleuze and Guattari*. MIT Press.

Massumi, B. (2002). *Parables for the virtual*. Duke University Press.

Mol, A., & Law, J. (1994). Regions, networks and fluids: Anaemia and social topology. *Social Studies of Science, 24*, 641–671.

Murcio, R., Masucci, A. P., Arcaute, E., & Batty, M. (2015). Multifractal to monofractal evolution of the London street network. *Physical Review E, 92*(6), 062130.

Paasi, A. (2011). Geography, space, and the reemergence of topological thinking. *Dialogues in Human Geography, 1*(3), 299–303.

Papadimitriou, F. (2020a). *Spatial complexity. Theory, mathematical methods and applications*. Springer.

Papadimitriou, F. (2020b). The topological basis of spatial complexity. In *Spatial complexity. Theory, mathematical methods and applications* (pp. 63–79). Springer.

Papadimitriou, F. (2020c). Geophilosophy and epistemology of spatial complexity. In *Spatial complexity. Theory, mathematical methods and applications* (pp. 263–278). Springer.

Phillips, J. W. P. (2013). On topology. *Theory, Culture & Society, 30*(1), 1–31.

Purevtseren, M., Tsegmid, B., Indra, M., & Sugar, M. (2018). The fractal geometry of urban land use: The case of Ulaanbaatar city, Mongolia. *Land, 7*(2), 67.

Richardson, L. F. (1961). The problem of contiguity: An appendix to statistics of deadly quarrels. General systems: Yearbook of the Society for the Advancement of systems theory. *The Society for General Systems Research, 6*(139), 139–187.

Rodin, V., & Rodina, E. (2000). The fractal dimension of Tokyo's streets. *Fractals, 8*(04), 413–418.

Rotman, B. (2012). Topology, algebra, diagrams. *Theory, Culture & Society, 29*(4/5), 247–260.

Santos, J. L., & Oliveira, A. (2021). Traveling salesman problem in a geographic information management system. In *Progress in industrial mathematics: Success stories: The industry and the academia points of view* (pp. 131–144). Springer.

Secor, A. (2013). 2012 urban geography plenary lecture topological city. *Urban Geography, 34*(4), 430–444.
Secor, A. J. (2013a). Topological city. *Urban Geography, 34*, 430–444.
Shields, R. (1997). Flow as a new paradigm. *Space and Culture, 1*, 1–7.
Shields, R. (2012). Cultural topology: The seven bridges of Konigsberg, 1736. *Theory, Culture & Society, 29*(4/5), 43–57.
Shields, R. (2013). *Spatial questions: Cultural topologies and social spatialisation*. Sage.
Sui, L., Wang, H., Wu, J., Zhang, J., Yu, J., Ma, X., & Sun, Q. (2022). Fractal description of rock fracture networks based on the space syntax metric. *Fractal and Fractional, 6*(7), 353.
Weinberger, D. (2003). Why open spectrum matters. The end of the broadcast nation. http://www.evident.com.
Xia, Y., Zhu, M., Gu, Q., Zhang, L., & Li, X. (2016). Toward solving the Steiner travelling salesman problem on urban road maps using the branch decomposition of graphs. *Information Sciences, 374*, 164–178.
Zhang, H., Lan, T., & Li, Z. (2022). Fractal evolution of urban street networks in form and structure: A case study of Hong Kong. *International Journal of Geographical Information Science, 36*(6), 1100–1118.

Part II
Geo-Topology in Geographical Analysis

Chapter 2
Geoinformatics and Topological Data Analysis

As for the mathematical magnitudes,
length is generated first, then width and last is depth

"τὰ μαθηματικὰ μεγέθη . . .
πρῶτον μὲν γὰρ ἐπὶ μῆκος γίγνεται,
εἶτα ἐπὶ πλάτος, τελευταῖον δ' εἰς βάθος"
(Aristotle, 384–322 b.C., "Metaphysics" 12, 1077$^{\alpha}$)

2.1 Simplexes, Boundaries and GIS

Sen (1976) was one of the first to focus on the relationships between topology and geography. Subsequently, using point-set topology, Beguin and Thisse (1979) defined "sets of places" that contain at least two places and suggested the use of the Lebesgue measure for the theoretical foundation of geographical space.

To GIS experts, it is customary to use "Triangulated Irregular Networks" (TINs) in vectorial representations of Digital Elevation Models. Behind the end-result of triangulation however, lie important topological properties. A triangulation of the geographical space is one of the operations that may result in "simplicial complexes". The elementary constituents of these "complexes" are "simplexes", that is algebraic-topological entities of dimensions 0 (points), 1 (segments), 2 (triangles), 3 (tetrahedra) etc. Given k points $x_1,\ldots,x_k$, the "smallest convex set" can be identified, in which a point x is defined from the set containing the linear combinations $x = \lambda_1 v_1 + \ldots + \lambda_k v_k$. where the λ_i are all nonnegative real numbers and $\lambda_1 + \ldots + \lambda_k = 1$. Such a set is a "simplex" of dimension k (or a k-simplex) and the points $x_1,\ldots,x_k$ are the vertices of the simplex. Simplexes fitted together form "complexes", which can be polyhedra. A finite collection of simplexes in a Euclidean space E^n is called a "simplicial complex" if, whenever a simplex lies in the collection, then so does each of its faces and, whenever two simplexes of the collection intersect, they do so with a face in common.

The use of notions of cell complexes and simplicial complexes for the description of 3D spatial data first appeared in the late 1980s–early 1990s (Carlson, 1987; Brisson, 1989; Molenaar, 1990; Pigot, 1991; Pilouk ct al., 1994). Following Egenhofer et al. (1989), the topology of the geographic space consists in a geometric level (purely spatial) and a non-spatial level, to which the concept of inheritance

© The Author(s), under exclusive license to Springer Nature Switzerland AG 2023

F. Papadimitriou, *Geo-Topology*, GeoJournal Library 133,
https://doi.org/10.1007/978-3-031-48185-7_2

applies (Cardelli, 1984), defining the direct connection of units at the geometric level with semantics at the non-spatial level. In the same way but with a different terminology, maps have been defined as sets of "tuples" (Scholl & Voisard, 1989), that is geometric regions with associated non-geometric information.

In the 1990s, a substantial part of the research in applications of topology to GIS theory focused on the definition and mathematical description of topological relations between spatial regions (Clementini & di Felice, 1997; Clementini et al., 1993, 1994; Godoy & Rodríguez, 2002; Ellul & Haklay, 2007; Billen & Kurata, 2008). These approaches extended from the formalization of the (most useful) 2D spatial relationships to point-set topology (Egenhofer & Franzosa, 1991), or even to cases where the genus of a surface changes (Egenhofer et al., 1994).

Major steps forward were made thereafter with alternative definitions of possible 3D topologies of spatial objects (Lienhardt, 1991; Billen & Zlatanova, 2003; Cardoze et al., 2006; Ellul & Haklay, 2006; de Almeida et al., 2007; Escobar-Molano et al., 2007; Breunig & Zlatanova, 2011; Mazroob Semnani et al., 2018) and thus, various models describing intersections among spatial regions appeared in the literature (Egenhofer & Shariff, 1998; Shariff et al., 1998; Chen et al., 2001; Nedas et al., 2007; Deng et al., 2007; Liu & Shi, 2007; Kurata, 2008; Egenhofer & Dube, 2009). Some of these ideas developed further towards algebraic models of topological relations (see i.e. Yuan et al., 2014; Yu et al., 2016).

Perhaps, the notion of "topological boundary" has been (and still is) the foremost important in all these areas of topological applications in geoinformatics. The boundary of a subset A of a topological space X is the set:

$$\partial A = Cl(A) - Int(A)$$

where $Cl(A)$ is the closure of A and $Int(A)$ is the interior of A. Thus, the interior and the boundary of a surface are disjoint. Also, if A is a subset of a topological space X and y a point in X, then $y \in \partial A$ if and only if every neighbourhood of y intersects both X-A and A. As an example, in the 1D topology of the real numbers, *let* $A = [-9,5]$. Then, $Cl(A) = [-9,5]$ and $Int(A) = (-9,5)$, and hence $\partial A = \{-9, 5\}$.

In general, if A is a subset of a topological space X, then the following statements about the boundary of A hold:

(i) ∂A is closed
(ii) $\partial A \cap Int(A) = \varnothing$
(iii) $\partial A \cup Int(A) = Cl(A)$
(iv) $\partial A \subset A$ if and only if A is closed
(v) $\partial A \cap A = \varnothing$ if and only if A is open
(vi) $\partial A = \varnothing$ if and only if A is both open and closed
(vii) $\partial A = Cl(A) \cap Cl(X - A)$

In 2D, we are interested in identifying the ways that two closed curves A and B (planar regions of R^2) may relate to one another (with A defining an area larger than B). The basic algebraic relationships between two regions in GIS or in maps can be defined by logical operators: intersection, union NOT, XOR that correspond to different topological relations (Fig. 2.1). The Open Geospatial Consortium has

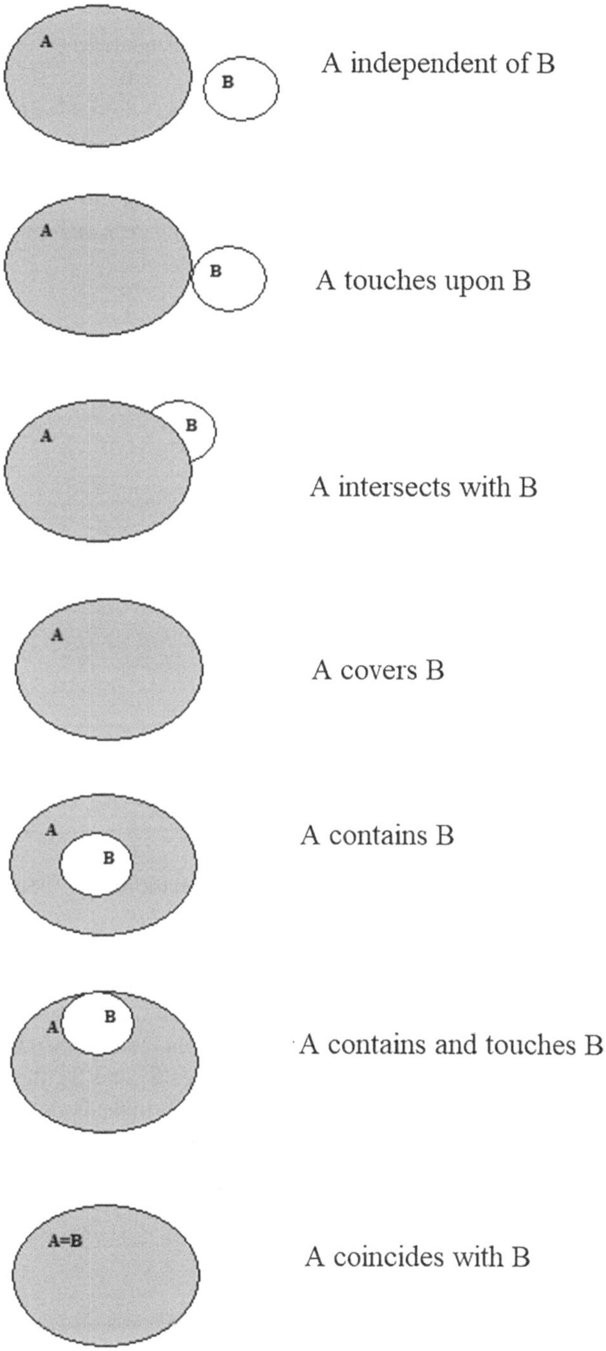

Fig. 2.1 Defining spatial relationships between two closed regions in R^2

Table 2.1 Possible relationships between two regions (*A* and *B*)

Relationship between *A* and *B*	Formulation
A equals *B*	$A = B$
A touches *B*	$A \cap B = \partial A \cap \partial B$
A is disjoint from *B*	$A \cap B \neq \varnothing$
A is within *B*	$A \subset B$
A contains *B*	$A \supset B$
A overlaps *B*	$IntA \cap IntB \neq \varnothing$, $A \not\subset B$, $B \not\subset A$

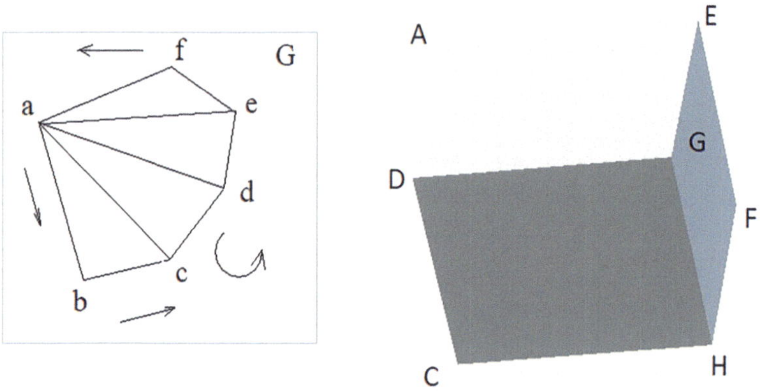

Fig. 2.2 An oriented curve G, defined by the points a,b,c,d,e,f that can be allocated on the surface of a sphere, enclosing four triangles (left) and calculation of the boundaries on a 3d surface, i.e. of a cube (right)

proposed a formulation covering the possible relationships between *A* and *B* (Table 2.1).

The boundary of a sum of simplexes is the sum of boundaries of the respective simplexes, and hence the boundary operator is additive: if C_1 and C_2 are simplexes, then $\partial\ (C_1 + C_2) = \partial\ (C_1) + \partial\ (C_2)$.

Also, there is a relationship between boundaries and homeomorphisms. If $h{:}S_1 \rightarrow S_2$ is a homeomorphism between the surfaces S_1 and S_2, then *h* maps the interior of S_1 to the interior of S_2, and the boundary of S_1 to the boundary of S_2. This can be seen in the example of the oriented curve G (Fig. 2.2).

If these triangles are oriented as indicated, their union is:

$$(a, b, c) + (a, c, d) + (a, d, e) + (a, e, f)$$

and the boundary of this union of triangles is:

$$\begin{aligned}
&\partial(\mathrm{a,b,c}) + \partial(\mathrm{a,c,d}) + \partial(\mathrm{a,d,e}) + \partial(\mathrm{a,e,f}) \\
&= (\mathrm{a,b}) + (\mathrm{b,c}) + (\mathrm{c,a}) + (\mathrm{a,c}) + (\mathrm{c,d}) + (\mathrm{d,a}) + (\mathrm{a,d}) + (\mathrm{d,e}) + (\mathrm{e,a}) + (\mathrm{a,e}) + (\mathrm{e,f}) + (\mathrm{f,a}) = \\
&= (\mathrm{a,b}) + (\mathrm{b,c}) + (\mathrm{c,a}) - (\mathrm{c,a}) + (\mathrm{c,d}) + (\mathrm{d,a}) - (\mathrm{d,a}) + (\mathrm{d,e}) + (\mathrm{e,a}) - (\mathrm{e,a}) + (\mathrm{e,f}) + (\mathrm{f,a}) = \\
&= (\mathrm{a,b}) + (\mathrm{b,c}) + (\mathrm{c,d}) + (\mathrm{d,e}) + (\mathrm{e,f}) + (\mathrm{f,a}) = \\
&= \mathrm{G}
\end{aligned}$$

Similarly for boundaries in 3D:

$$\partial(\mathrm{ABCDEFGH}) = (\mathrm{ABCD}) + (\mathrm{DAEG}) + (\mathrm{GHFE}) + (\mathrm{EFBC}) + \ldots$$

Following Egenhofer's DE9IM (dimensionally-extended nine intersection matrix), if the interior of a 2D spatial region *A is* A^0, its exterior A^- and its boundary ∂A, then these are related by the matrix" (DE9IM) model:

$$\mathrm{DE9IM} = \begin{bmatrix} \dim(A^0 \cap B^0) & \dim(A^0 \cap \partial B) & \dim(A^0 \cap B^-) \\ \dim(\partial A \cap B^0) & \dim(\partial A \cap \partial B) & \dim(\partial A \cap B^-) \\ \dim(A^- \cap B^0) & \dim(A^- \cap \partial B) & \dim(A^- \cap B^-) \end{bmatrix} = \\ = \begin{bmatrix} 2 & 1 & 2 \\ 1 & 0 & 1 \\ 2 & 1 & 2 \end{bmatrix}$$

The resulting matrix on the right-hand side of the equation shows the dimension of each intersection. Thus, the intersection of two interiors is a planar region so its dimension is 2, the intersection of the boundary of an area and the exterior of another is a line or polygon, and hence its dimension is 1. Similarly, the intersection of two boundaries is a point (dimension 0) and so on for all the entries of the matrix. Further elaborations of this model have produced, i.e. an expansion to 27-intersections (Shen et al., 2017), while other systems of classification of topological relationships have been proposed, by taking into account closeness (Egenhofer & Shariff, 1998; Shariff et al., 1998; Nedas et al., 2007; Dube et al., 2015) and metric differences (Egenhofer & Dube, 2009; Sridhar et al., 2011). Addressing the need for standardization in the domain of spatial boundaries among spatial objects, the Open Geospatial Conrortium suggested GeoSPARQL as a standard for the identification of topological relations (OGC, 2010), which has also been implemented in spatiotemporal RDF (Kyzirakos et al., 2012).

Besides boundaries, simplexes and simplicial complexes, another central concept of topology that is important for geoinformatics is triangulations of spaces. These are often useful in 3D spatial analysis and depend on some simplicial complex *K* under consideration and on a triangulating homeomorphism h (a triangulation of a topological space X consists of a simplicial complex K and a homeomorphism $h: K \rightarrow X$). Hence, a *topological* space is "triangulable", if it is homeomorphic to the union of a finite collection of simplexes in a euclidean space. A triangulable space can be derived from a set of triangles by identifying the vertexes and edges of these triangles, subject to the restriction that any two triangles are either identified along an edge or at a vertex, or are disjoint (two triangles in a triangulable space have either two common vertexes, or one, or none). A sphere can be decomposed in tetrahedra (since it is homeomorphic to them) and the tetrahedra in triangles and the "Triangulated Irregular Network" is thus a set of triangles producing a 3D discrete topography of a geographical space.

2.2 Topological Data Analysis and Vietoris-Rips Complexes

The set of all intersections of neighborhoods around each point of a space defines the "topological base" of that space. A topology T of a space X is the sum total $\Omega(X)$ of all the topological neighborhoods of X that conform to the "Hausdorff axioms" which require that any unions and finite overlaps correspond to a neighborhood that also belongs to $\Omega(X)$ and that the empty set and the full set X belong to $\Omega(X)$ as well. Consequently, to create the topology on X, it is necessary to create the topological base B first, which is composed from the set of basic neighborhoods that, if combined, produce any neighborhood of $\Omega(X)$. Much like the union of neighborhoods belongs to $\Omega(X)$, any union of bases produces a base that also belongs to the same topology. From an initial set of neighborhoods u_1,u_2,u_3,u_4,u_5 of a space X and their intersections $(u_{12},u_{14},u_{24},u_{13},u_{15},u_{35},u_{124},u_{135})$, it is possible to define the topology $\Omega(X)$ (Fig. 2.3).

At this point open up two avenues for future research in geographical analysis at the intersection of topology and geography. The first is derived from the translation of topology into an algebraic representation by creating partially-ordered sets ("posets") or "Hasse diagrams" (Fig. 2.4). Hasse diagrams have proven useful to analyze land use and landscapes with the aim of deriving priorities for land management (Papadimitriou, 2012).

The second is to single out the "*Vietoris-Rips*" simplicial complex N_T which summarizes the key properties of the topology T. This is a simplicial complex (Fig. 2.5) in which every neighborhood is represented by a node (dimension 0) of a graph, all neighborhood intersections by lines (dimension 1), intersections of three neighborhoods by triangles (dimension 2) and those among four neighborhoods by tetrahedra. When two neighborhoods u_i and u_j intersect, the intersection u_{ij} corresponds to a line and when three neighborhoods intersect (i.e. u_i, u_j, u_k) they are represented by a triangle u_{ijk}.

A Vietoris-Rips complex is derived from a map of points by progressively expanding radii around each and all points which correspond to neighborhoods. Expectedly, the resulting simplicial complex becomes increasingly larger with increasing radii. Consequently, the *Betti numbers* can be calculated for each value

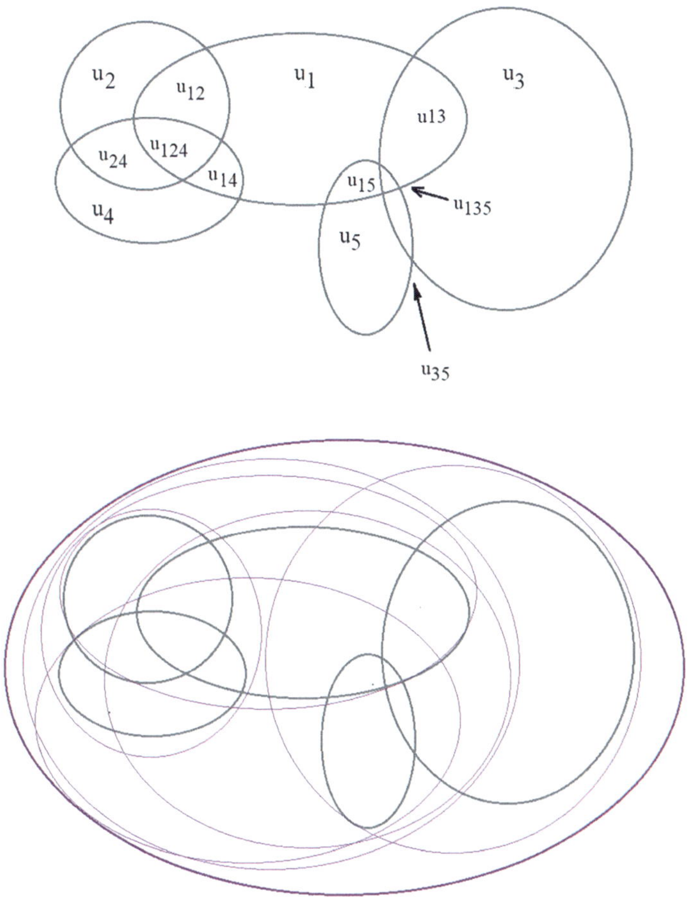

Fig. 2.3 A set of neighborhoods and their intersections (above) leads to a topology (below) that is defined by all the unions of all neighborhoods and their overlaps

Fig. 2.4 Construction of the Hasse diagram of neighborhoods. Nodes correspond to neighborhoods and links to their intersections

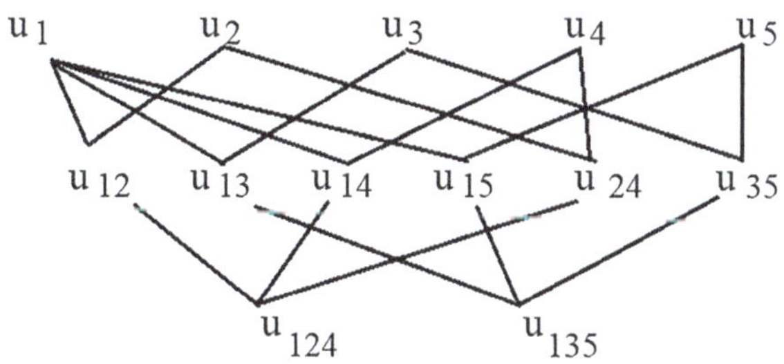

of the radius *r*. The first Betti number (b_0) stands for the number of separate (disjoint but internally connected) entities, the second number (b_1) for the holes and the third (b_2) for the internal cavities (Fig. 2.6). The procedure is called "Vietoris-Rips

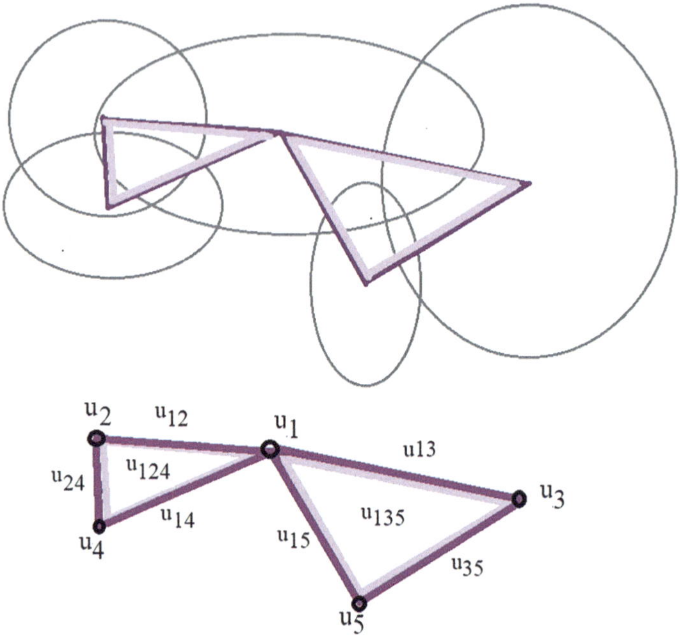

Fig. 2.5 Construction of a Vietoris-Rips complex from neighborhoods. Nodes correspond to neighborhoods, lines to intersections of two neighborhoods and triangle areas to intersections of three neighborhoods (see text for explanation)

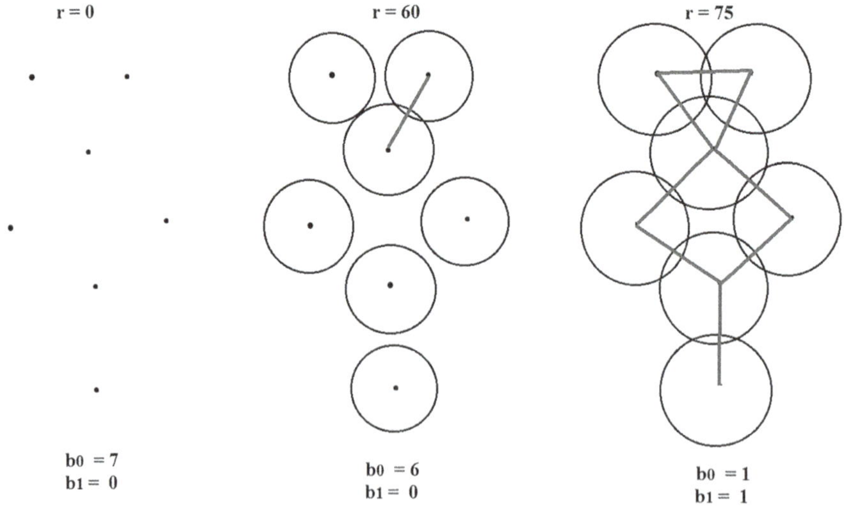

Fig. 2.6 An example of the Vietoris-Rips filtration procedure and construction of the Vietoris-Rips complex from an initial set of points on the plane. The values of radius r are calculated in pixels in this case, with b_0 and b_1 the Betti numbers

filtration" and is part of the "Topological Data Analysis" techniques that have been developed (Zomorodian, 2010, 2012; Epstein et al., 2011; Bubenik, 2015; Taylor et al., 2015; Munch, 2017; Tierny, 2017; Wasserman, 2018; Atienza et al., 2019; Asao et al., 2021; Carlsson & Vejdemo-Johansson, 2021; Chazal & Michel, 2021). With the Vietoris-Rips complexes, point data dispersed in the geographical space accept topological classifications and if "buffer zones" surround these points, the analysis of the expansion of those buffer zones can be based on methods of topological data analysis.

References

Asao, Y., Nagase, J., Sakamoto, R., & Takagi, S. (2021). Image recognition via Vietoris-Rips complex. *arXiv preprint arXiv:*2109.02231.

Atienza, N., Gonzalez-Diaz, R., & Rucco, M. (2019). Persistent entropy for separating topological features from noise in vietoris-rips complexes. *Journal of Intelligent Information Systems, 52*, 637–655.

Beguin, H., & Thisse, J. F. (1979). An axiomatic approach to geographic space. *Geographical Analysis, 11*(4), 325–341.

Billen, R., & Kurata, Y. (2008). Refining topological relations between regions considering their shapes. In M. Raunbal, J. Miller, A. U. Frank, et al. (Eds.), *Geographic information science* (Lecture notes in computer science) (pp. 18–32). Springer.

Billen, R., & Zlatanova, S. (2003). 3D spatial relationship model: A useful concept for 3D cadastre? *Computers, Environment and Urban Systems, 27*, 411–425.

Breunig, M., & Zlatanova, S. (2011). 3D geo-database research: Retrospective and future directions. *Computers and Geosciences, 37*(7), 791–803.

Brisson, E. (1989). Representing geometric structures in d dimensions: Topology and order. In *Proceedings of the fifth annual symposium on computational geometry SCG'89* (pp. 218–227). ACM.

Bubenik, P. (2015). Statistical topological data analysis using persistence landscapes. *Journal of Machine Learning Research, 16*(1), 77–102.

Cardelli, L. (1984). A semantics of multiple inheritance. In *International symposium on semantics of data types* (pp. 51–67). Springer.

Cardoze, D., Miller, G., & Phillips, T. (2006, July 26–28). Representing topological structures using cell-chains. In *Proceedings of the geometric modeling and processing—GMP 2006*. Kim, M.-S., Shimada, K., Eds). (pp. 248 266). : Springer.

Carlson, E. (1987). Three dimensional conceptual modeling of subsurface structures. *Technical Papers of ASPRS/ACSM Annual Convention, 4*, 188–200.

Carlsson, G., & Vejdemo-Johansson, M. (2021). *Topological data analysis with applications*. Cambridge University Press.

Chazal, F., & Michel, B. (2021). An introduction to topological data analysis: Fundamental and practical aspects for data scientists. *Frontiers in Artificial Intelligence, 4*, 108.

Chen, J., Li, C., Li, Z., & Gold, C. (2001). A Voronoi-based 9-intersection model for spatial relations. *International Journal of Geographical Information Science, 15*(3), 201–220.

Clementini, E., & di Felice, P. (1997). Approximate topological relations. *International Journal of Approximate Reasoning, 16*, 173–204.

Clementini, E., di Felice, P., & van Oosterom, P. J. M. (1993). A small set of formal topological relations suitable for end-user interaction. In *Proceedings of the 3th international symposium on large spatial databases* (pp. 277–295). Springer.

Clementini, E., Sharma, J., & Egenhofer, M. J. (1994). Modelling topological spatial relations: Strategies for query processing. *Computer Graphics, 18*(6), 815–822.

de Almeida, J. P., Morley, J. G., & Dowman, I. J. (2007). Graph theory in higher order topological analysis of urban scenes. *Computers, Environment and Urban Systems, 31*, 426–440.

Deng, M., Cheng, T., Chen, X., & Li, Z. (2007). Multi-level topological relations between spatial regions based upon topological invariants. *GeoInformatica, 11*, 239–267.

Dube, M.P., Barrett, J.V., & Egenhofer, M.J. (2015, October 12–16). From metric to topology: Determining relations in discrete space. In Spatial information theory: 12th international conference, COSIT 2015, Proceedings 12 (pp. 151–171). : Springer.

Egenhofer, M. J., & Dube, M. P. (2009). Topological relations from metric refinements. In *Proceedings of the 17th ACM SIGSPATIAL international conference on advances in geographic information systems* (pp. 158–167).

Egenhofer, M. J., & Franzosa, R. D. (1991). Point-set topological spatial relations. *International Journal of Geographical Information Systems, 5*, 161–174.

Egenhofer, M. J., & Shariff, A. R. B. (1998). Metric details for natural-language spatial relations. *ACM Transactions on Information Systems (TOIS), 16*(4), 295–321.

Egenhofer, M. J., Frank, A. U., & Jackson, J. P. (1989). A topological data model for spatial databases. *Springer Lecture Notes in Computer Science, 7649*, 271–286.

Egenhofer, M. J., Clementini, E., & di Felice, P. (1994). Topological relations between regions with holes. *International Journal of Geographical Information Systems, 8*(2), 129–144.

Ellul, C., & Haklay, M. (2006). Requirements for topology in 3D GIS. *Transactions in GIS, 10*, 157–175.

Ellul, C., & Haklay, M. (2007). The research agenda for topological and spatial databases. *Computers, Environment and Urban Systems, 31*, 373–378.

Epstein, C., Carlsson, G., & Edelsbrunner, H. (2011). Topological data analysis. *Inverse Problems, 27*(12), 120201.

Escobar-Molano, M. L., Barret, D. A., Carson, E., et al. (2007). A representation for databases of 3D objects. *Computers, Environment and Urban Systems, 31*, 409–425.

Godoy, F., & Rodríguez, A. (2002). A quantitative description of spatial configurations. In *Advances in spatial data handling: 10th international symposium on spatial data handling* (pp. 299–311). Springer.

Kurata, Y. (2008, September 23–26). The 9+−intersection: A universal framework for modeling topological relations. In *Geographic information science: 5th international conference, GIScience 2008, Park City, UT, USA*. Proceedings 5 (pp. 181–198). : Springer.

Kyzirakos, K., Karpathiotakis, M., & Koubarakis, M. (2012). Strabon: A semantic geospatial DBMS. In P. Cudré-Mauroux, J. Heflin, E. Sirin, T. Tudorache, J. Euzenat, M. Hauswirth, J. X. Parreira, J. Hendler, G. Schreiber, A. Bernstein, & E. Blomqvist (Eds.), *ISWC 2012, Boston, USA, Springer lecture notes in computer science, 7649* (pp. 295–311).

Lienhardt, P. (1991). Topological models for boundary representation: A comparison with *n*-dimensional generalized maps. *Computer-Aided Design, 23*, 59–82.

Liu, K., & Shi, W. (2007). Extended model of topological relations between spatial objects in geographic information systems. *International Journal of Applied Earth Observation and Geoinformation, 9*(3), 264–275.

Mazroob Semnani, N., Kuper, P. V., Breunig, M., & Al-Doori, M. (2018). Towards an intelligent platform for big 3d geospatial data management. *ISPRS Annals of the Photogrammetry, Remote Sensing and Spatial Information Sciences, 4*, 133–140.

Molenaar, M. (1990). A formal data structure for three dimensional GIS in geographic information systems. In *Proceedings of 4th international symposium on spatial data handling, 2* (pp. 830–843).

Munch, E. (2017). A user's guide to topological data analysis. *Journal of Learning Analytics, 4*(2), 47–61.

Nedas, K. A., Egenhofer, M. J., & Wilmsen, D. (2007). Metric details of topological line–line relations. *International Journal of Geographical Information Science, 21*(1), 21–48.

O'Rourke, J. (1985). Finding minimal enclosing boxes. *International Journal of Computer & Information Sciences, 14*(3), 183–199.
OGC (2010). GeoSPARQL – A geographic query language for RDF data.
Papadimitriou, F. (2012). Modelling landscape complexity for land use management in Rio de Janeiro, Brazil. *Land Use Policy, 29*(4), 855–861.
Pigot, S. (1991). Topological models for 3D spatial information systems. In *Proceedings of the AutoCarto conference, Baltimore, MD, vol.6* (pp. 368–392). ASPRS (American Society for Photogrammetry and Remote Sensing).
Pilouk, M., Tempfli, K., & Molenaar, M. (1994). A tetrahedron-based 3D vector data model for geo-information. In M. Molenaar & S. de Hoop (Eds.), *AGDM'94 spatial data modelling and query languages for 2D and 3D applications* (pp. 129–140). Netherlands Geodetic Comm.
Scholl, M., & Voisard, A. (1989). Thematic map modelling: Design and implementation of large spatial databases. *Springer Lecture Notes in Computer Science, 409*, 167–190.
Sen, A. (1976). On a class of map transformations. *Geographical Analysis, 8*, 23–37.
Shariff, A. R. B., Egenhofer, M. J., & Mark, D. M. (1998). Natural-language spatial relations between linear and areal objects: The topology and metric of English-language terms. *International Journal of Geographical Information Science, 12*(3), 215–245.
Shen, J., Zhou, T., & Chen, M. (2017). A 27-intersection model for representing detailed topological relations between spatial objects in two-dimensional space. *ISPRS International Journal of Geo-Information, 6*(2), 37.
Sridhar, M., Cohn, A. G., & Hogg, D. C. (2011). From video to RCC8: exploiting a distance based semantics to stabilise the interpretation of mereotopological relations. In *Spatial Information Theory: 10th International Conference, COSIT 2011, Belfast, ME, USA, September 12–16, 2011.* Proceedings 10 (pp. 110–125). Springer.
Taylor, D., Klimm, F., Harrington, H. A., Kramár, M., Mischaikow, K., Porter, M. A., & Mucha, P. J. (2015). Topological data analysis of contagion maps for examining spreading processes on networks. *Nature Communications, 6*(1), 7723.
Tierny, J. (2017). *Topological data analysis for scientific visualization* (Vol. 3). Springer.
Wasserman, L. (2018). Topological data analysis. *Annual Review of Statistics and Its Application, 5*, 501–532.
Yu, Z., Luo, W., Yuan, L., Hu, Y., Zhu, A. X., & Lü, G. (2016). Geometric algebra model for geometry-oriented topological relation computation. *Transactions in GIS, 20*(2), 259–279.
Yuan, L., Yu, Z., Luo, W., Yi, L., & Lü, G. (2014). Multidimensional-unified topological relations computation: A hierarchical geometric algebra-based approach. *International Journal of Geographical Information Science, 28*(12), 2435–2455.
Zomorodian, A. (2010). Fast construction of the Vietoris-rips complex. *Computers & Graphics, 34*(3), 263–271.
Zomorodian, A. (2012). Topological data analysis. *Advances in Applied and Computational topology, 70*, 1–39.

Chapter 3
Geo-topology of Landscape Boundaries

A boundary is not that at which something stops,
but, as the Greeks recognized,
the boundary is that from which something begins its presence.
(Martin Heidegger, 1889–1976, "Building, dwelling, thinking")

3.1 Boundaries and the Jordan Curve Theorem

The adoption of topological methods in spatial analysis is often not a matter of choice, but the only way to analyse spatial data. Calculating the area that results from boundary change on a planar region R may explain why this is so. In general, a region R has an area A equal to:

$$A = \iint_R \sqrt{1 + p^2 + q^2} dxdy$$

where

$$p = \frac{\partial f(x, y)}{\partial x} \text{ and } q = \frac{\partial f(x, y)}{\partial y}$$

Using calculus to evaluate the change of area in R in the course of time is not the least complicated approach. This is because the curve of a boundary is given by the parametric equation

$$\vec{r} = (x(u, t), y(u, t)),$$

so the area inscribed within the boundary (which may as well be curly) is defined as:

© The Author(s), under exclusive license to Springer Nature Switzerland AG 2023

F. Papadimitriou, *Geo-Topology*, GeoJournal Library 133,
https://doi.org/10.1007/978-3-031-48185-7_3

$$A = \int_0^1 x(u,t)\frac{\partial y(u,t)}{\partial u}du.$$

This quantity can be calculated only as long as the points (x,y) are known. But they are known only as long as the parametric equations $x(u,t)$ and $y(u,t)$ are also known.

An alternative is to resort to vectors and curvature, in which case the formula of area change is

$$\frac{dA(t)}{dt} = \int K \cdot V_n \cdot ds$$

where K is the curvature of the boundary, V_n is the normal velocity of the movement of the boundary and S is the arc length, with

$$V_n = \frac{\Delta d}{\Delta t} = \left(\frac{\vec{r}(s,t,t+\Delta t) - \vec{r}(s,t)}{\Delta t}\right) \cdot \vec{n},$$

$$\vec{n} = \begin{pmatrix} 0 & 1 \\ -1 & 0 \end{pmatrix} \cdot \vec{T}$$

and $\vec{T}$ the tangent vector.

Evidently, none of these analytic approaches is applicable to maps that are (most likely) replete with many and spatially complex boundaries. This leaves us with topological approaches (i.e. based on rasterization) as the only practically applicable alternative for the calculation of areas resulting from boundary changes. Furthermore, analytic methods are inadequate to provide information about relationships among areas for the entire map (whereas topological methods using networks of boundaries can) and this becomes explicit when a geographical study entails high complexity (Papadimitriou, 2020a, b, c).

With these considerations, the next step is to focus on certain key notions of topology that are related to boundaries. The first one relates to openness and the number of internal boundaries of a planar region.

Let X be a topological space and $x \in X$. An open set B containing x is a neighborhood of x. Now let X be a topological space and A subset of X. The set A is "open" in X if and only if for each $x \in X$, there is a neighborhood B of x, such that $x \in B \subset A$. The intersection of open sets is also an open set. Open sets can be constructed from within any topology and their number depends on how *"fine"* a topology is. If X is a set with T_1 and T_2 are two topologies on X, and if $T_1 \subset T_2$, then T_2 is *"finer"* than T_1, and T_1 is *"coarser"* than T_2. Further, a topological space X is a *Hausdorff* space if for every pair of distinct points in X, there exists a pair of disjoint open sets, with each containing one of the points. Or, more formally put, a topological space X is Hausdorff, if for every pair of distinct points a and b in X, there exist disjoint neighborhoods N_a and N_b of a and b, respectively. A non-Hausdorff topological space can not be induced by a metric. But if X is a metrizable topological space and Y is a space homeomorphic to X, then Y is also metrizable. Whether a space is metrizable or not, can be verified by using the "Urysohn Metrization Theorem".

A surface is "*closed*" if it is compact, connected and without a boundary. Otherwise stated, a surface is closed if it is a compact and connected Hausdorff space in which each point has a neighborhood homeomorphic to the plane.

Now let there be two spaces X and Y and a function f: $X \rightarrow Y$. This function is an *embedding of* X in Y if it maps the space X homeomorphically to a subspace $f(X)$ in Y.

Arcs and simple closed curves can be embedded in the 2D space and knots are embeddings of the circle in the 3D sphere.

If Y is a topological space and if f: $[-1,1] \rightarrow Y$ is an embedding, then the image of the map f is an *arc* in Y. In the same way, if $f{:}\Gamma \rightarrow Y$ is an embedding, then the image of f is a *simple closed curve* Γ in Y.

In 1887, Camille Jordan proved a difficult yet obvious statement: that a closed curve on the plane separates the plane into the union of three disjoint sets: the curve itself, the interior of the curve and the area outside the curve. Formally, let Γ be a simple closed curve in R^2. Then R^2-Γ consists of two components and Γ is the boundary of each one of these two components in R^2. Hence, any path joining a point in the interior with the outside will necessarily intersect the curve, while any two points in the interior of the curve can be joined by a path that lies entirely in the interior. The same applies to the area outside of the curve (any two points that lie on the outside can be joined by a continuous path). This seeming as a self-evident facy can be misleading, as soon as it is considered that proving it for all possible closed curves, even for curves of infinite length is very difficult (Boltyanskii & Efremovich, 2001). Yet, there is the underlying assumption that the curve should lie on a surface of genus zero. For if it is a closed curve wrapped around the hole of a torus, then it does not divide the space on the surface of the torus in two parts, because it can not be decided which part of the torus lies "inside" and what is "outside" of that curve (Fig. 3.1).

A closed curve has the interesting property that any point that lies inside it is connected with any point outside the curve with an odd number of intersection points. If the point lies outside the closed curve, then any straight line connecting it with another outside point, the boundary of the closed curve will cross at an even number of points (Fig. 3.2).

In geography, curves that define boundaries dividing the inside from the outside are encountered among countries (in which case they are called borders) or among their administrative divisions and there is a large literature that pertains to inclusion/exclusion relationships in the geographical space (Portugali, 1984; Taylor, 1999; Schnell & Yoav, 2001; Drever, 2004; Clark & Clark, 2009; Cloutier-Fisher & Kobayashi, 2009; Krivo et al., 2013; Martin, 2015; Papageorgiou, 2018), i.e. relating to "insularity" and isolation from the outside world (Bennett et al., 2018; Shen, 2019; Baldacchino & Starc, 2021). Further, boundaries may not separate physical spaces only, but perceived "fragmented spaces" also (Mondada, 2010; Gélinas-Lemaire, 2018; Öner, 2020). Sacred spaces in particular, are spaces apart

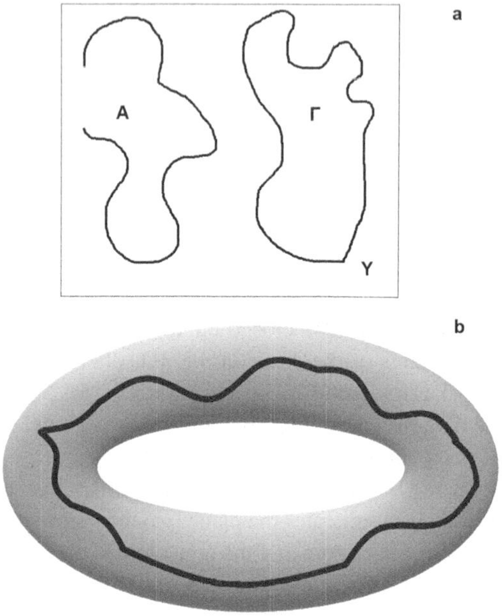

Fig. 3.1 Whereas an arc and a simple closed curve can always be defined in R^2 (**a**), a closed curve on the surface of a torus may not divide the surface of the torus (**b**) in inside and outside regions

that are defined by cultural and physical boundaries (Grapard, 1982; Jackson & Henrie, 1983; Lane, 2002; Hamilton, 2003; Harvey, 2006; Stump, 2008; Knott, 2009; Dora, 2018). However, the notion of boundary is in no other field of geography more frequently examined than in landscape geography.

3.2 Landscape Boundaries

Indeed, landscape boundaries constitute probably one of the most important topological features of landscape analysis (Wiens et al., 1985; Johnston et al., 1992; Forman & Moore, 1992; Forman, 1995; Metzger & Muller, 1996; Kent et al., 1997; Fortin et al., 2000; Csillag & Kabos, 2002; Koen et al., 2010; Martín-López et al., 2017). As often the case in many geographical settings is, landscape boundaries are neither static nor clearly defined. For instance, they may become increasingly fluid and may depend on the scale of observation or on the levels of accuracy and/or precision by which they are mapped out (Fig. 3.3).

In landscape ecology, the concept of "ecotone" is characteristic in this respect: ecotones are ecological transition zones, gradients between two vegetation zones, often associated with some environment-induced stress (Kolasa & Zalewski, 1995; Smith et al., 1997; Allen & Breshears, 1998; Lloyd et al., 2000; Heasley, 2003; Kark & Van Rensburg, 2006; Holland, 2012; Della Dora, 2012). A concept akin to the ecotone is the ecocline, that is gradient zone with relatively heterogeneous but environmentally stable characteristics (van der Maarel, 1990). It has to be noticed

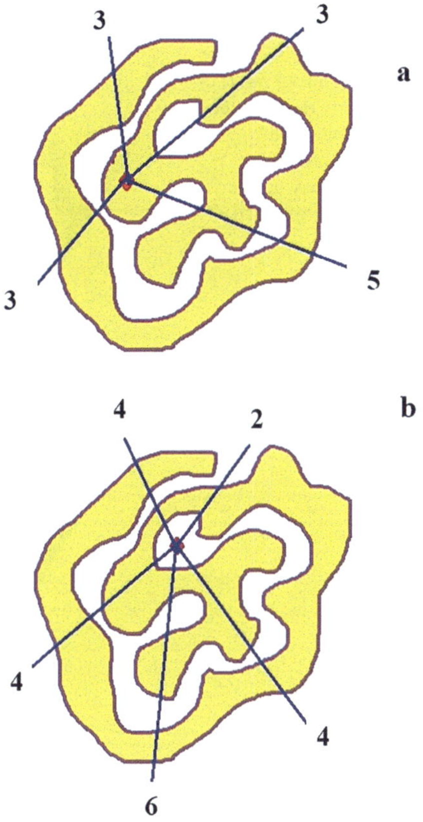

Fig. 3.2 A point inside a closed curve on the plane is connected to a point outside the curve by an odd number of times of intersections with the boundaries of the closed curve (**a**). If the point lies outside the curve (**b**), then there is an even number of intersections. Numbers show the cardinalities of intersections along each segment

however that, under certain conditions, drawing boundaries in the geographical space may not make sense (i.e. if there is high richness of species or diversity within a region and the individuals present a high dispersion within that region).

Landscape fragmentation is a process that leads to increasing the number of landscape boundaries and is a central theme in landscape ecology, landscape analysis and landscape planning (King & Burton, 1982; Bentley, 1987; Gardner et al., 1993; Jaeger, 2000; Bogaert et al., 2000; Nagendra et al., 2004; Munroe et al., 2005; Zeng & Wu, 2005; Gonzalez-Abraham et al., 2007; Moser et al., 2007; Irwin & Bockstael, 2007; Girvetz et al., 2008; Dewan et al., 2012; Farley et al., 2012; Marcantonio et al., 2013; Demetriou et al., 2013; Fan & Myint, 2014; Li & Yang, 2015; Alencar et al., 2015; Jaeger et al., 2016; De Montis et al., 2017; Rosa et al., 2017; De Montis et al., 2020; Lawrence et al., 2021;Wang, 2022). The core idea is that more landscape boundaries among patches indicate landscape fragmentation. Equivalently, a lower number of boundaries (or a lower number of boundary lengths) may be indicative of landscape homogenization (Fig. 3.4), which is generally undesirable (Papadimitriou & Mairota, 1998a, b).

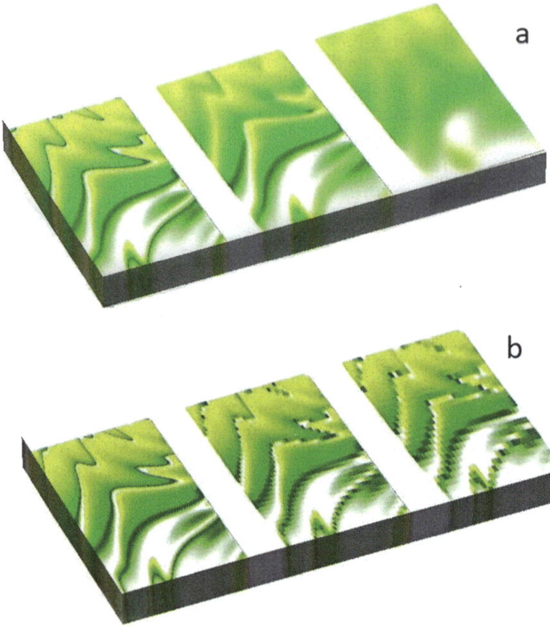

Fig. 3.3 Some boundaries change in the course of time (**a**), becoming either more fluid (from left to right) or more clearly defined (the reverse), but digital representations of boundaries may appear coarser, i.e. zigzag (**b**)

Fig. 3.4 Landscape fragmentation (left part) vs. landscape homogenization (right part)

Hence, interlacing of various natural landscape vegetation covers (i.e. forest and shrubland) should be preferred with the aim of maintaining and even enhancing ecological interactions among patches in landscape planning (Fig. 3.5).

But it often is a question of balance: whereas interlacing between two different natural vegetation types is preferable, the complexity of interlacing should not lead to thinning boundaries to the extent that they will eventually lead to dissociation of pieces of land from their initial patch. However, it is ecologically preferable to occasionally have larger borders between two natural vegetation types instead of a single enclosed area (Fig. 3.6).

Another situation that is often encountered is when some landscape type that surrounds another one by an outer ring "eats out" the internal landscape type

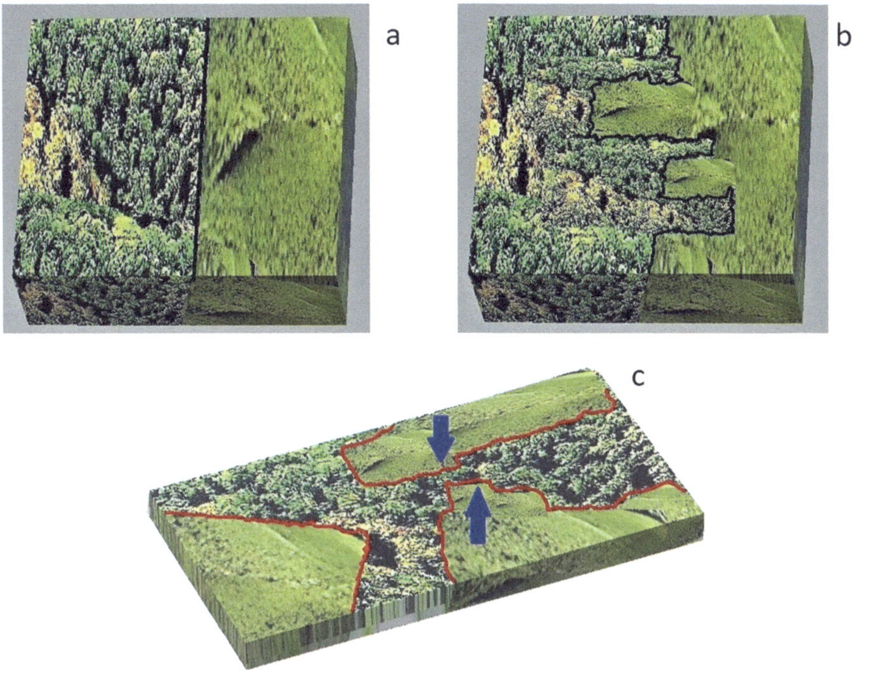

Fig. 3.5 Clear-cut boundaries between two landscape types that both consist of natural vegetation (forest and shrubland here) should be avoided in landscape planning (**a**). Interlacing is preferable instead (**b**), although it should not lead to excessive thinning of boundaries that might eventually result in isolated patches (**c**)

Fig. 3.6 It is preferable to increase the length of the boundary between two natural vegetation types so as to enhance the communication of species (**b**), instead of having single enclosed areas (**a**)

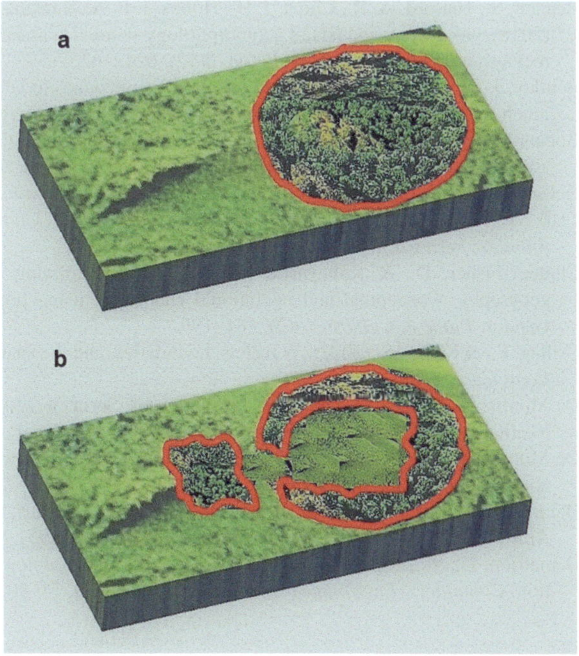

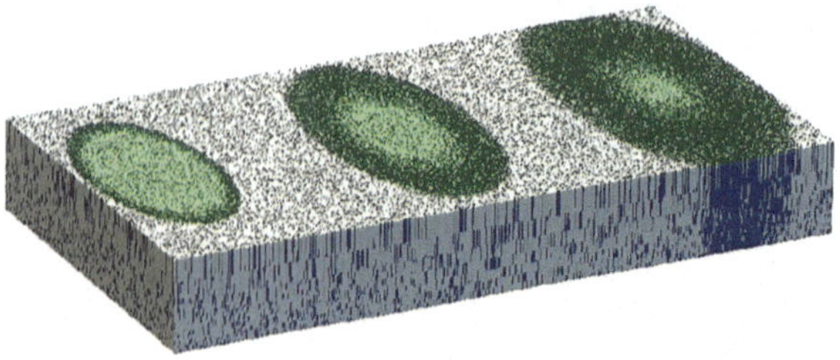

Fig. 3.7 Landscape boundary dynamics are often generated from the interactions between an "outer" and an "inner" landscape type

(Fig. 3.7) and progressively dominates it. The reverse can also happen: the internal landscape type may progressively expand and absorb its surrounding. A typical form of the latter case is the expansion of urban sprawls at the detriment of their surrounding forests or other types of natural vegetation.

References

Alencar, A. A., Brando, P. M., Asner, G. P., & Putz, F. E. (2015). Landscape fragmentation, severe drought, and the new Amazon forest fire regime. *Ecological Applications, 25*(6), 1493–1505.

Allen, C. D., & Breshears, D. D. (1998). Drought-induced shift of a forest–Woodland ecotone: Rapid landscape response to climate variation. *Proceedings of the National Academy of Sciences, 95*(25), 14839–14842.

Baldacchino, G., & Starc, N. (2021). The virtues of insularity: Pondering a new chapter in the historical geography of islands. *Geography Compass, 15*(12), e12596.

Bennett, S., Halford, A. R., Choat, J. H., Hobbs, J. P. A., Santana-Garcon, J., Ayling, A. M., et al. (2018). Geography and Island geomorphology shape fish assemblage structure on isolated coral reef systems. *Ecology and Evolution, 8*(12), 6242–6252.

Bentley, J. W. (1987). Economic and ecological approaches to land fragmentation: In defense of a much-maligned phenomenon. *Annual Review of Anthropology, 16*(1), 31–67.

Bogaert, J., Van Hecke, P., Eysenrode, D. S. V., & Impens, I. (2000). Landscape fragmentation assessment using a single measure. *Wildlife Society Bulletin, 64*, 875–881.

Boltyanskii, V. G., & Efremovich, V. A. (2001). *Intuitive combinatorial topology*. Springer.

Clark, E., & Clark, T. L. (2009). Isolating connections–Connecting isolations. *Geografiska Annaler: Series B, Human Geography, 91*(4), 311–323.

Cloutier-Fisher, D., & Kobayashi, K. M. (2009). Examining social isolation by gender and geography: Conceptual and operational challenges using population health data in Canada. *Gender, Place & Culture, 16*(2), 181–199.

Csillag, F., & Kabos, S. (2002). Wavelets, boundaries, and the spatial analysis of landscape pattern. *Ecoscience, 9*(2), 177–190.

De Montis, A., Martín, B., Ortega, E., Ledda, A., & Serra, V. (2017). Landscape fragmentation in Mediterranean Europe: A comparative approach. *Land Use Policy, 64*, 83–94.

De Montis, A., Serra, V., Ganciu, A., & Ledda, A. (2020). Assessing landscape fragmentation: A composite indicator. *Sustainability, 12*(22), 9632.

Della Dora, V. (2012). Setting and blurring boundaries: Pilgrims, tourists, and landscape in Mount Athos and Meteora. *Annals of Tourism Research, 39*(2), 951–974.

Demetriou, D., Stillwell, J., & See, L. (2013). A new methodology for measuring land fragmentation. *Computers, Environment and Urban Systems, 39*, 71–80.

Dewan, A. M., Yamaguchi, Y., & Ziaur Rahman, M. (2012). Dynamics of land use/cover changes and the analysis of landscape fragmentation in Dhaka metropolitan, Bangladesh. *Geo Journal, 77*, 315–330.

Dora, V. D. (2018). Infrasecular geographies: Making, unmaking and remaking sacred space. *Progress in Human Geography, 42*(1), 44–71.

Drever, A. I. (2004). Separate spaces, separate outcomes? Neighbourhood impacts on minorities in Germany. *Urban Studies, 41*(8), 1423–1439.

Fan, C., & Myint, S. (2014). A comparison of spatial autocorrelation indices and landscape metrics in measuring urban landscape fragmentation. *Landscape and Urban Planning, 121*, 117–128.

Farley, K. A., Ojeda-Revah, L., Atkinson, E. E., & Eaton-González, B. R. (2012). Changes in land use, land tenure, and landscape fragmentation in the Tijuana River watershed following reform of the ejido sector. *Land Use Policy, 29*(1), 187–197.

Forman, R. T. (1995). Some general principles of landscape and regional ecology. *Landscape Ecology, 10*, 133–142.

Forman, R. T., & Moore, P. N. (1992). Theoretical foundations for understanding boundaries in landscape mosaics. In A. J. Hansen (Ed.), *Landscape boundaries: Consequences for biotic diversity and ecological flows* (pp. 236–258). Springer-Verlag.

Fortin, M. J., Olson, R. J., Ferson, S., Iverson, L., Hunsaker, C., Edwards, G., et al. (2000). Issues related to the detection of boundaries. *Landscape Ecology, 15*, 453–466.

Gardner, R. H., O'Neill, R. V., & Turner, M. G. (1993). *Ecological implications of landscape fragmentation* (pp. 208–226). Springer.

Gélinas-Lemaire, V. (2018). The narrative lives of places: Literature as an architecture. *Contemporary French and Francophone Studies, 22*(4), 454–462.

Girvetz, E. H., Thorne, J. H., Berry, A. M., & Jaeger, J. A. (2008). Integration of landscape fragmentation analysis into regional planning: A statewide multi-scale case study from California, USA. *Landscape and Urban Planning, 86*(3–4), 205–218.

Gonzalez-Abraham, C. E., Radeloff, V. C., Hammer, R. B., Hawbaker, T. J., Stewart, S. I., & Clayton, M. K. (2007). Building patterns and landscape fragmentation in northern Wisconsin, USA. *Landscape Ecology, 22*, 217–230.

Grapard, A. G. (1982). Flying mountains and walkers of emptiness: Toward a definition of sacred space in Japanese religions. *History of Religions, 21*(3), 195–221.

Hamilton, J. F. (2003). Psychological geography and sacred space in Sand's La petite Fadette. In *Geo/graphies* (pp. 87–98). Brill.

Harvey, T. (2006). Sacred spaces, common places: The cemetery in the contemporary American city. *Geographical Review, 96*(2), 295–312.

Heasley, L. (2003). Shifting boundaries on a Wisconsin landscape: Can GIS help historians tell a complicated story? *Human Ecology, 31*, 183–213.

Holland, M. (Ed.). (2012). *Ecotones: The role of landscape boundaries in the management and restoration of changing environments*. Springer.

Irwin, E. G., & Bockstael, N. E. (2007). The evolution of urban sprawl: Evidence of spatial heterogeneity and increasing land fragmentation. *Proceedings of the National Academy of Sciences, 104*(52), 20672–20677.

Jackson, R. H., & Henrie, R. (1983). Perception of sacred space. *Journal of Cultural Geography, 3*(2), 94–107.

Jaeger, J. A. (2000). Landscape division, splitting index, and effective mesh size: New measures of landscape fragmentation. *Landscape Ecology, 15*, 115–130.

Jaeger, J. A., Soukup, T., Schwick, C., Madriñán, L. F., & Kienast, F. (2016). Landscape fragmentation in Europe. In *European landscape dynamics* (pp. 187–228). CRC Press.

Johnston, C. A., Pastor, J., & Pinay, G. (1992). Quantitative methods for studying landscape boundaries. In A. J. Hansen & F. Di Castri (Eds.), *Landscape boundaries: Consequences for biotic diversity and ecological flows* (pp. 107–125). Springer-Verlag.

Kark, S., & Van Rensburg, B. J. (2006). Ecotones: Marginal or central areas of transition? *Israel Journal of Ecology & Evolution, 52*(1), 29–53.

Kent, M., Gill, W. J., Weaver, R. E., & Armitage, R. P. (1997). Landscape and plant community boundaries in biogeography. *Progress in Physical Geography, 21*(3), 315–353.

King, R., & Burton, S. (1982). Land fragmentation: Notes on a fundamental rural spatial problem. *Progress in Human Geography, 6*(4), 475–494.

Knott, K. (2009). Geography, space and the sacred. In *The Routledge companion to the study of religion* (pp. 490–505). Routledge.

Koen, E. L., Garroway, C. J., Wilson, P. J., & Bowman, J. (2010). The effect of map boundary on estimates of landscape resistance to animal movement. *PLoS One, 5*(7), e11785.

Kolasa, J., & Zalewski, M. (1995). Notes on ecotone attributes and functions. *Hydrobiologia, 303*, 1–7.

Krivo, L. J., Washington, H. M., Peterson, R. D., Browning, C. R., Calder, C. A., & Kwan, M. P. (2013). Social isolation of disadvantage and advantage: The reproduction of inequality in urban space. *Social Forces, 92*(1), 141–164.

Lane, B. C. (2002). *Landscapes of the sacred: Geography and narrative in American spirituality*. Johns Hopkins University Press.

Lawrence, A., Friedrich, F., & Beierkuhnlein, C. (2021). Landscape fragmentation of the Natura 2000 network and its surrounding areas. *PLoS One, 16*(10), e0258615.

Li, S., & Yang, B. (2015). Introducing a new method for assessing spatially explicit processes of landscape fragmentation. *Ecological Indicators, 56*, 116–124.

Lloyd, K. M., McQueen, A. A., Lee, B. J., Wilson, R. C., Walker, S., & Wilson, J. B. (2000). Evidence on ecotone concepts from switch, environmental and anthropogenic ecotones. *Journal of Vegetation Science, 11*(6), 903–910.

Marcantonio, M., Rocchini, D., Geri, F., Bacaro, G., & Amici, V. (2013). Biodiversity, roads, & landscape fragmentation: Two Mediterranean cases. *Applied Geography, 42*, 63–72.

Martin, D. (2015). From spaces of exception to 'campscapes': Palestinian refugee camps and informal settlements in Beirut. *Political Geography, 44*, 9–18.

Martín-López, B., Palomo, I., García-Llorente, M., Iniesta-Arandia, I., Castro, A. J., Del Amo, D. G., et al. (2017). Delineating boundaries of social-ecological systems for landscape planning: A comprehensive spatial approach. *Land Use Policy, 66*, 90–104.

Metzger, J. P., & Muller, E. (1996). Characterizing the complexity of landscape boundaries by remote sensing. *Landscape Ecology, 11*, 65–77.

Mondada, L. (2010). Reassembling fragmented geographies. In M. Büscher, J. Urry, & K. Witchger (Eds.), *Mobile methods* (pp. 138–163). Routledge.

Moser, B., Jaeger, J. A., Tappeiner, U., Tasser, E., & Eiselt, B. (2007). Modification of the effective mesh size for measuring landscape fragmentation to solve the boundary problem. *Landscape Ecology, 22*, 447–459.

Munroe, D. K., Croissant, C., & York, A. M. (2005). Land use policy and landscape fragmentation in an urbanizing region: Assessing the impact of zoning. *Applied Geography, 25*(2), 121–141.

Nagendra, H., Munroe, D. K., & Southworth, J. (2004). From pattern to process: Landscape fragmentation and the analysis of land use/land cover change. *Agriculture, Ecosystems & Environment, 101*(2–3), 111–115.

Öner, I. M. (2020). *Isolated spaces, fragmented places. Spatial literary studies: Interdisciplinary approaches to space, geography, and the imagination* (p. 126). Routledge.

Papadimitriou, F. (2020a). *Spatial complexity. Theory, mathematical methods and applications*. Springer.

Papadimitriou, F. (2020b). The topological basis of spatial complexity. In *Spatial complexity. Theory, mathematical methods and applications* (pp. 63–79). Springer.

Papadimitriou, F. (2020c). The algorithmic basis of spatial complexity. In *Spatial complexity. Theory, mathematical methods and applications* (pp. 81–99). Springer.

Papadimitriou, F., & Mairota, P. (1998a). Land use diversity. In P. Mairota, J. Thornes, & N. Geeson (Eds.), *Atlas of Mediterranean environments in Europe*. J. Wiley.

Papadimitriou, F., & Mairota, P. (1998b). Agriculture. In P. Mairota, J. Thornes, & N. Geeson (Eds.), *Atlas of Mediterranean environments in Europe*. J. Wiley.

Papageorgiou, Y. (2018). *The isolated city state: An economic geography of urban spatial structure* (Vol. 18). Routledge.

Portugali, J. (1984). On relevance in geography: Thünen's isolated state in relation to agriculture and political economy. *Geoforum, 15*(2), 201–207.

Rosa, I. M., Gabriel, C., & Carreiras, J. M. (2017). Spatial and temporal dimensions of landscape fragmentation across the Brazilian Amazon. *Regional Environmental Change, 17*, 1687–1699.

Schnell, I., & Yoav, B. (2001). The sociospatial isolation of agents in everyday life spaces as an aspect of segregation. *Annals of the Association of American Geographers, 91*(4), 622–636.

Shen, Y. (2019). Segregation through space: A scope of the flow-based spatial interaction model. *Journal of Transport Geography, 76*, 10–23.

Smith, T. B., Wayne, R. K., Girman, D. J., & Bruford, M. W. (1997). A role for ecotones in generating rainforest biodiversity. *Science, 276*(5320), 1855–1857.

Stump, R. W. (2008). *The geography of religion: Faith, place, and space*. Rowman & Littlefield Publishers.

Taylor, P. J. (1999). Places, spaces and Macy's: Place–space tensions in the political geography of modernities. *Progress in Human Geography, 23*(1), 7–26.

van der Maarel, E. (1990). Ecotones and ecoclines are different. *Journal of Vegetation Science, 1*, 135–138.

Wang, X. (2022). Changes in cultivated land loss and landscape fragmentation in China from 2000 to 2020. *Land, 11*(5), 684.

Wiens, J. A., Crawford, C. S., & Gosz, J. R. (1985). Boundary dynamics: A conceptual framework for studying landscape ecosystems. *Oikos, 45*(3), 421–427.

Zeng, H., & Wu, X. B. (2005). Utilities of edge-based metrics for studying landscape fragmentation. *Computers, Environment and Urban Systems, 29*(2), 159–178.

Chapter 4
Geo-topology of Networks of Borders

Geography blended with time equals destiny
(Joseph Brodsky, 1940–1996, "Strophes")

4.1 Geographical Networks

There are various types of geographical networks and topology has been instrumental in their analysis, also in tandem with non-topological spatial analyses. For instance, the topological analysis of urban street networks has been a recurrent theme in geographical research (Khasnabish, 1989; Jiang & Claramunt, 2004; Jiang, 2007; Jiang & Liu, 2009; Jiang et al., 2009; Hillier et al., 2010; Lin & Ban, 2013; Jiang et al., 2014; Lin & Ban, 2017; Lee & Jung, 2018; Kirkley et al., 2018; Sharifi, 2019). From some of these explorations of network structure and function, it has been possible to compare the topological properties of networks with other interesting features such as entropy (Boeing, 2019; Coutrot et al., 2022) or self-organization (Buhl et al., 2006), as exemplified by studies of street networks of London and Beijing (Masucci et al., 2009, 2014; Wang, 2015).

Aside of street networks, topological methods are also applicable to the analysis of infrastructure networks (power grids, sewer networks, gas distribution networks, telecommunications infrastructure networks). The topology of such networks is of utmost importance for their optimal design (Liu et al., 2011; Saldarriaga et al., 2021), for identifying their critical links and nodes (Svendsen & Wolthusen, 2007; Dunn & Wilkinson, 2013; Liu & Song, 2020) and for improving their controllability and robustness (Wang et al., 2013, 2016). Besides design, such network analyses may also have theoretical importance, i.e. comparing the growth of random networks (Schultz et al., 2014) or their co-evolution with other urban networks (Zischg et al., 2019). For such topological analyses of geographical networks, some known indices of network analysis are described next.

© The Author(s), under exclusive license to Springer Nature Switzerland AG 2023

F. Papadimitriou, *Geo-Topology*, GeoJournal Library 133,
https://doi.org/10.1007/978-3-031-48185-7_4

The *alpha* index (α) is defined as

$$\alpha = u/(2g - 5)$$

where u is the number of disconnected components of the network's graph. The values of alpha range from 0 (no circuits present) to 100% (a completely interconnected network). A similar index is the "*chromatic number*":

$$\mu = L\text{–}g + p.$$

The *beta* index (β) is defined as the ratio of edges (L) to vertices (g):

$$\beta = L/g$$

and the *gamma* index (γ), the prominent measure of network connectivity, is defined as the number of edges in a network divided by the maximum number of edges possible:

$$\gamma = \frac{2L}{g(g-1)}$$

where L is the number of links and g is the number of nodes. This index is also referred to as the "density" of a graph. For instance, a graph with a density equal to 0.2 means that it has several unlinked nodes (implying that it is not a very communicative network, as it has only 20% of the density of the interconnectedness that it might possibly have).

The "nodal degree" is another characteristic of the graph and the "diameter" is its major geodesic distance (the shortest distance between the two remotest nodes of the network measured by the minimum number of links in between them).

Beyond the geography of transport networks and infrastructures, a major issue that emerged along with the expansion of globalization and inter-regional trade and migration has been the rising significance of international and cross-border networks of people, goods, energy and information. As a matter of fact, cross-border activities have been on the rise during the last decades and relate to various cross-border complexities in (among other sectors) tourism (Gelbman & Timothy, 2011), healthcare (Kumar & Rodrigues, 2009), border markets (Lauritsen, 2012), landscapes (Papadimitriou, 2020), languages (Rash, 2002; Dunlop, 2013), education (Ford et al., 2009), illegal trade (Aguiar, 2012), cross-border conflicts (Lal, 2006), energy networks (Watcharejyothin & Shrestha, 2009; Bednarczyk et al., 2010), infrastructure projects (Hammons, 2010), taxation (Smith, 1999), power trade (Lama, 2000) etc. Of particular geographical interest are the environmental and ecological dimensions of cross-border land management. The European Union for instance, has fostered a pan-european architecture of ecological networks of protected areas (Natura 2000, Habitats Directive, Council Directive 92/43/EEC of 21.05.1992), establishing the connectivity of cross-border habitats across the EU territory. The "Natura 2000" network of protected sites connects approximately 26,000 sites that cover almost 1,000,000 km^2 and according to articles 3 and 10 of the Directive, the "ecological coherence" had to be fostered by these networks across the European territory (Opermanis et al., 2012; Sarvasová et al., 2012).

4.2 Network Topologies of Borders

Land borders are defined by physical geography (river basins, lakes, mountains etc), as well as by historical agreements and treaties among countries. Although, in some instances there may be some uncertainty whether the term "border" or "frontier" is more appropriate to use, we increasingly more talk about "borders" whereas the interest for borders is on the rise in the last years. For this reason, it was proposed that we may need a discipline/science/theory focusing on the analysis, planning and management of borders and cross-border areas, a "Synoriology" (from the greek word "Συνοριολογία" (Συν-οριο-λογία, /Syn-orio-logy is a composite word derived from the greek syn/συν = plus and orio/όριο = border, frontier, limit and logy/λόγος = scientific discourse), to denote the scientific theory of borders, frontiers, limits, boundaries (Papadimitriou, 2015).

Perhaps, one of the best ways to illustrate the power of topology in the analysis of borders is to explore topologically the changes in the borders of Europe during the twentieth century. Ending the "30-years war" in Europe, the treaty of Westfalia established borders on the basis of nation-states in 1648. In the modern era, the "Central Commission for the Navigation of the Rhine" (Commission Centrale pour la Navigation du Rhin, Zentralkommission für die Rheinschifffahrt Centrale Commissie voor de Rijnvaart, in French, German and Dutch respectively) has been regarded as the first effective cross-border cooperation in modern Europe. But the history of border changes in Europe is quite long. Tracing back the changes in the topology of borders among European countries from 1900 until 2022, it is possible to derive graph models (Figs. 4.1 and 4.2) with countries represented by nodes and borders among them by links (land borders have been taken into account only).

A first observation is that the numbers of borders and countries both increased in the course of this (longer than a century) time interval (Fig. 4.3). The values of two typical indices of network analysis (the average betweenness centrality and the average power centrality) reflect, in different ways, how central a network's nodes are (on the average) as they also increased during the same time interval (Fig. 4.4).

According to the radial representations of these networks (Figs. 4.4 and 4.5), each concentric circle has one degree higher per node. The outer circle has degree zero so the most linked countries (with borders) are those found at the center of the radial graph. These radial graphs reveal that the topology of the borders of the European countries gradually became more complex, with higher degrees per node (Fig. 4.5).

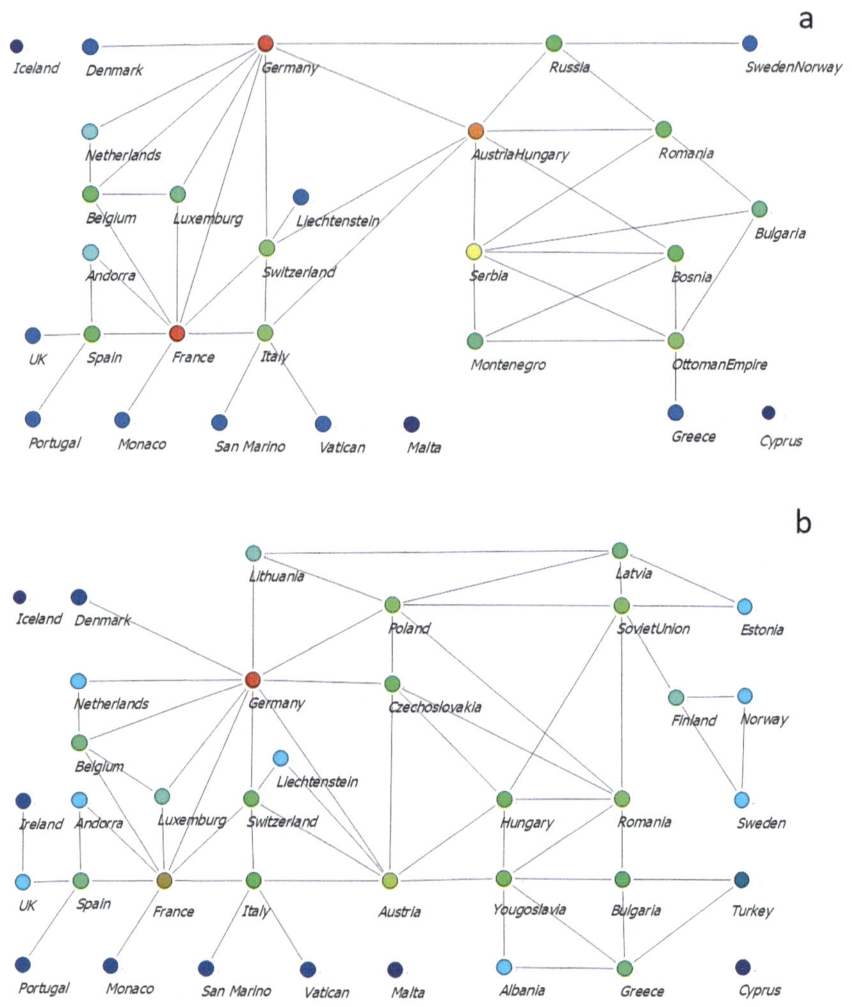

Fig. 4.1 The borders of European countries in 1900 (a) and 1936 (b)

The mean degree per node, the network spatial entropy and the mean distance increased from 1900 to 2022 (Fig. 4.6), while the degree density decreased.

The values of the α-index and β-index also reflect the increase in borders from 1900 until 2022 (Fig. 4.7).

The diameter of these networks increased by one at every subsequent year (7 in 1900, 8 in 1936, 9 in 1990 and 10 in 2022), implying that more borders should be crossed (on the average) if one needed to travel i.e. across Europe from end to end.

The topological analysis of the European borders suggests that the topology of the borders of the European countries has become increasingly more complex during the twentieth and the early twenty-first centuries and with increasing spatial entropy. It is

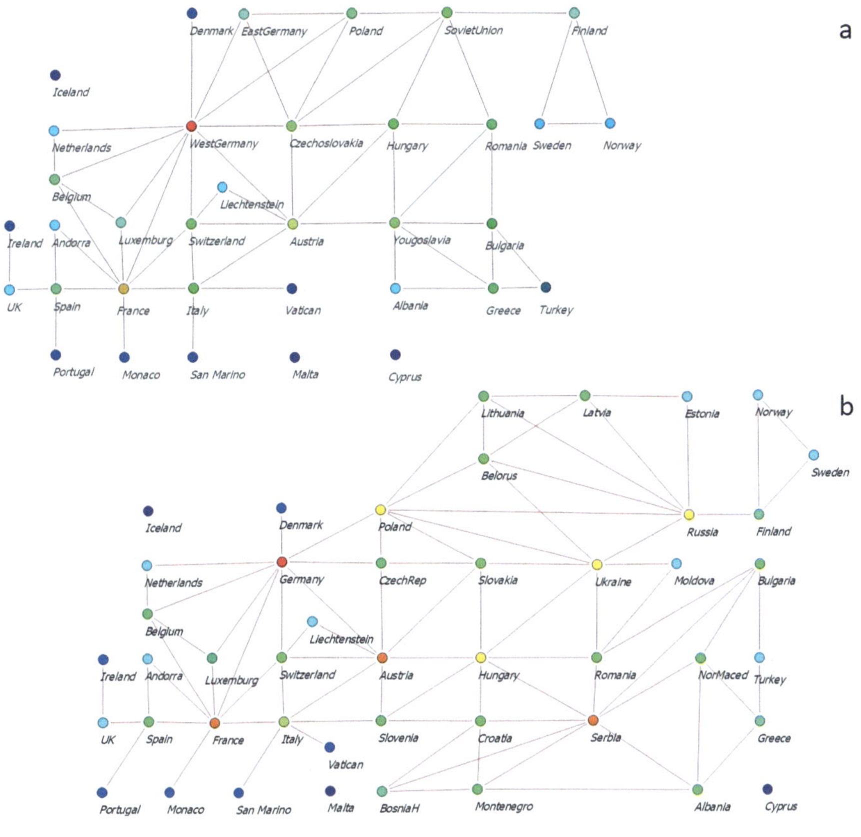

Fig. 4.2 The borders of European countries in 1990 (a) before the dissolution of the Soviet Union, Czechoslovakia and Yugoslavia and in 2022 (b)

natural to ask what would happen if all these trends persisted and lasted for one generation (although there may be several good reasons why these trends will *not* be sustained due to wars and break ups of countries, unifications of countries etc): what might be expected for 2050 (if the 1900–2022 trends continued unhampered) is a Europe with 74 countries with mean degree per country equal to 4 (Fig. 4.8).

In representing networks with graphs, issues of planarity may be important for visualization. Following the Robertson-Seymour theorem, a graph is planar if and only if it does not contain any graphs of the type "K_5" or "$K_{3,3}$", because the presence of anyone of these two graphs is evidence of *non-planarity* (Fig. 4.9a, b). These graphs can nevertheless be embedded in other surfaces without intersection. Since the graph $K_{3,3}$ can be arranged as sets of links joining two sets of three nodes each, it is interesting to explore its geographical significance: no three locations can be serviced by three different services without these services intersecting one another.

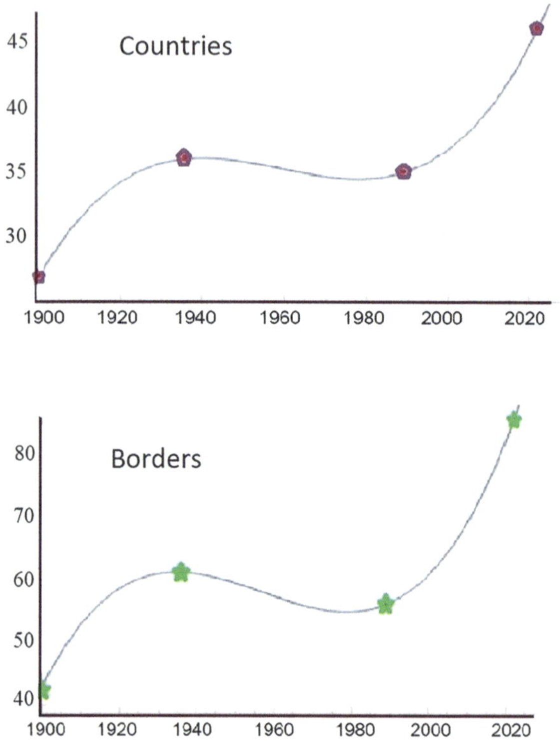

Fig. 4.3 The increase in the number of borders follows the increase in the number of countries (1900–2022) and they can be modeled by cubic polynomials (with $r = 1$ in both cases)

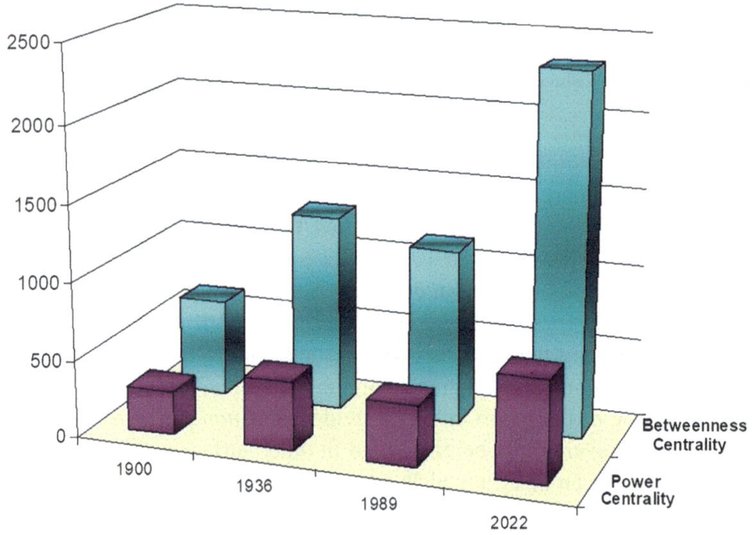

Fig. 4.4 The betweenness and power centralities of the networks of borders of European countries increased from 1900 until 2022

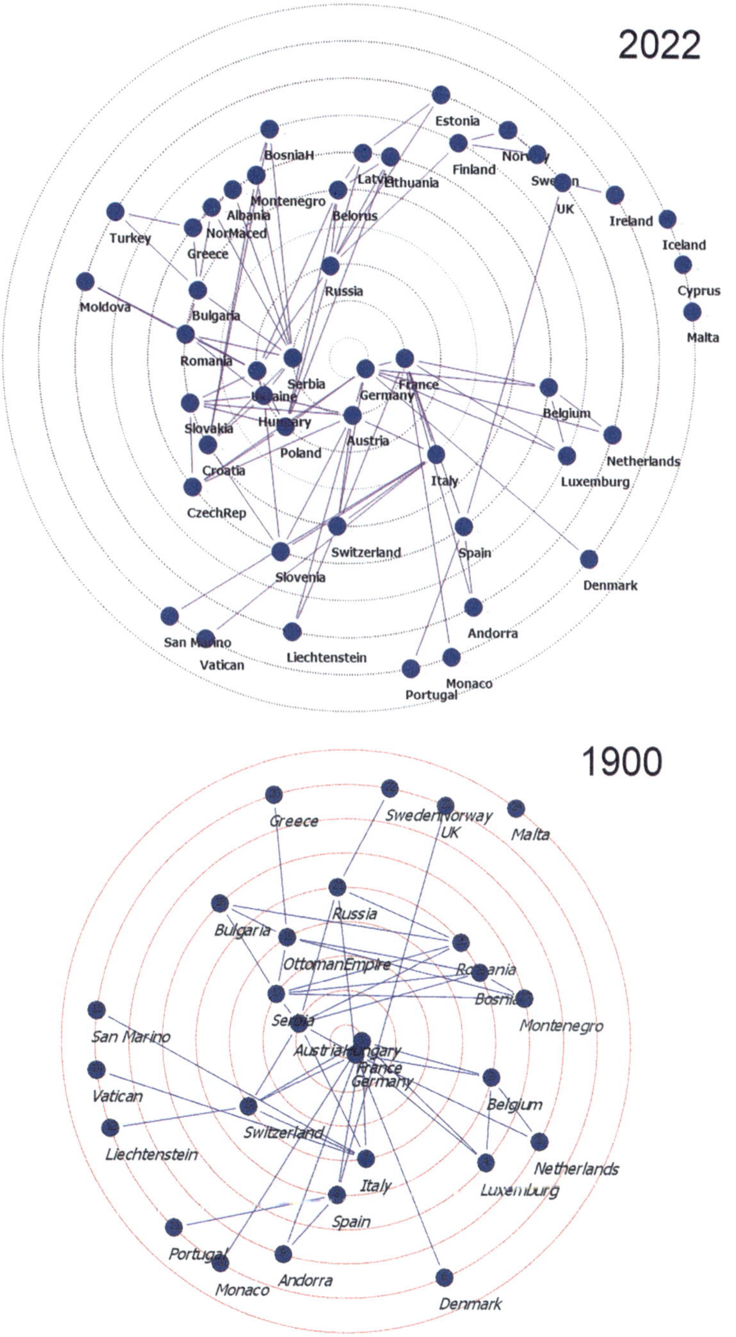

Fig. 4.5 The radial representations of the changing topologies of the borders of European countries (1900–2022) display increasingly larger radii as more countries appear and (expectedly) higher degrees per node

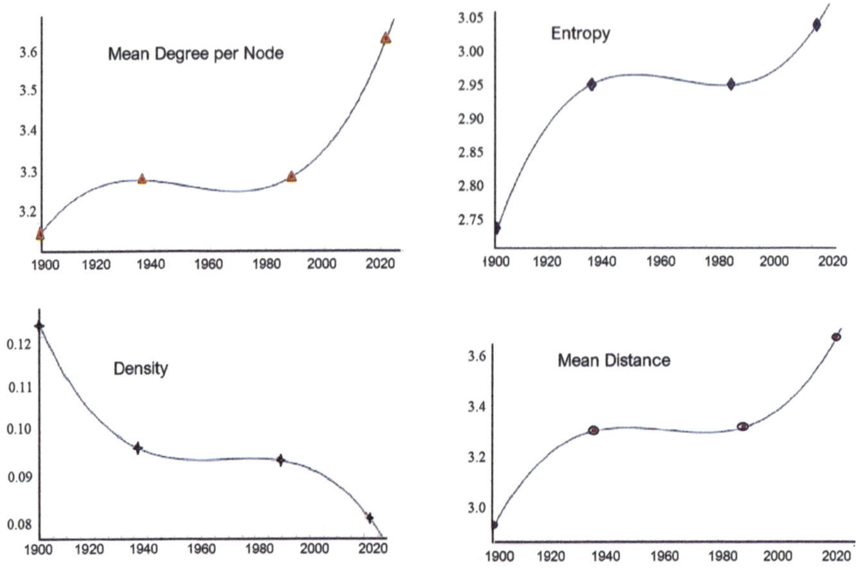

Fig. 4.6 Values of mean degree per node, density, entropy and mean distance and their cubic polynomial models ($r = 1$ in all cases)

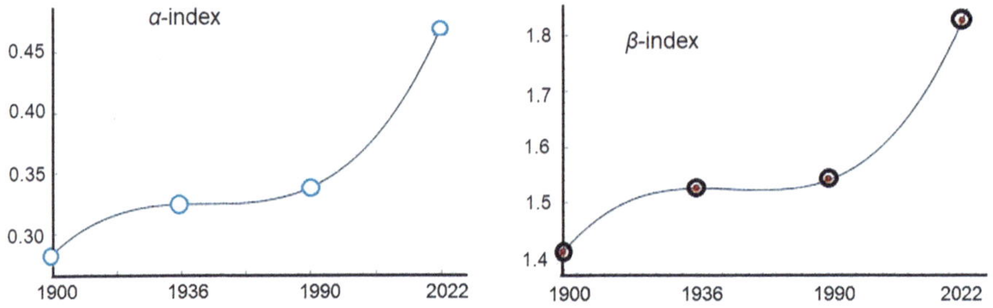

Fig. 4.7 Values of the α and β indexes and their cubic polynomial models ($r = 1$ in all cases)

Further, before any topological analysis of networks by using graphs is carried out, a decision has to be made about the countries that share multiple borders. If two countries are regarded as sharing one common border only, that implies a different geotopological analysis of their borders compared to considering them as having two different borders (Fig. 4.9c, d) and the corresponding graph models differ.

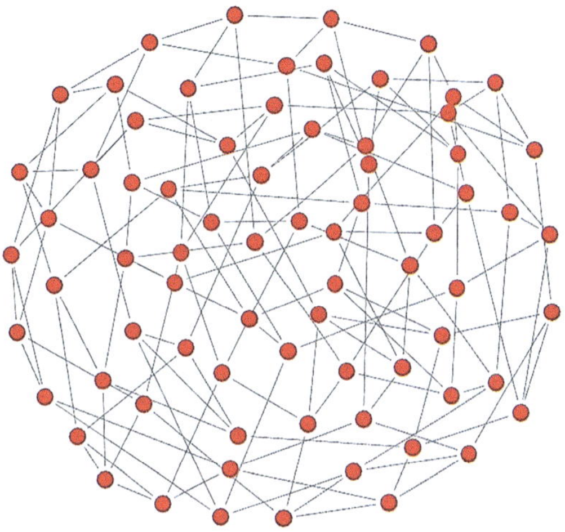

Fig. 4.8 A "Kamada-Kawai" representation of a random network of the topological structure of the countries-and-borders in Europe in 2050 (74 countries with mean degree per node equal to 4) if the 1900–2022 trends continued unhampered

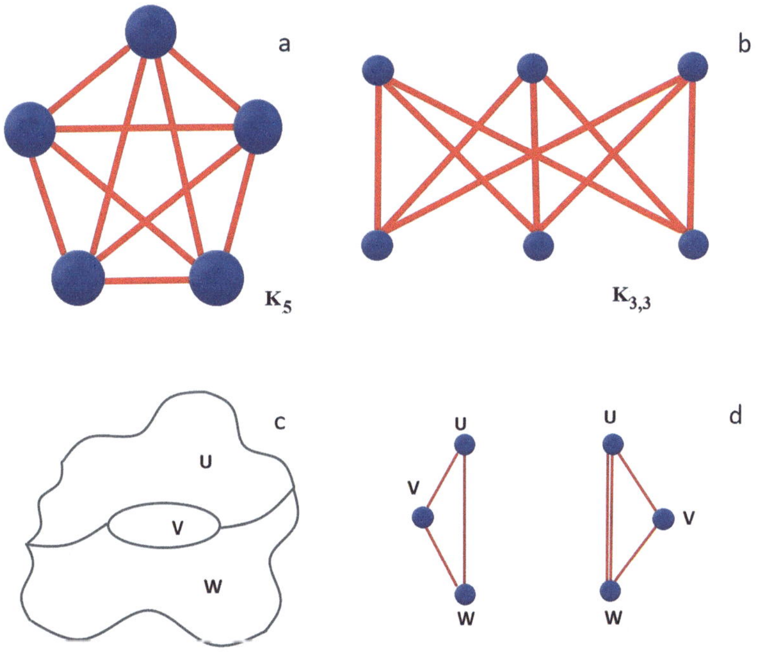

Fig. 4.9 The graphs K_5 and $K_{3,3}$ are two elementary non-planar graphs, which means that their links can not be drawn on the plane without intersections (a, b). Two different geotopological analyses apply depending on whether countries U and W are considered as sharing one common border (c) or two different borders (d) and so the corresponding graphs are also different in the two cases

References

Aguiar, J. C. G. (2012). Cities on edge: Smuggling and neoliberal policies at the Iguazú triangle. *Singapore Journal of Tropical Geography, 33*(2), 171–183.

Bednarczyk, T. P., Schiereck, D., & Walter, H. N. (2010). Cross-border acquisitions and shareholder wealth: Evidence from the energy and industry in central and Eastern Europe. *Journal for East European Management Studies, 15*(2), 106–127.

Boeing, G. (2019). Urban spatial order: Street network orientation, configuration, and entropy. *Applied Network Science, 4*(1), 1–19.

Buhl, J., Gautrais, J., Reeves, N., Solé, R. V., Valverde, S., Kuntz, P., & Theraulaz, G. (2006). Topological patterns in street networks of self-organized urban settlements. *The European Physical Journal B-Condensed Matter and Complex Systems, 49*, 513–522.

Coutrot, A., Manley, E., Goodroe, S., Gahnstrom, C., Filomena, G., Yesiltepe, D., et al. (2022). Entropy of city street networks linked to future spatial navigation ability. *Nature, 604*(7904), 104–110.

Dunlop, C. T. (2013). Mapping a new kind of European boundary: The language border between modern France and Germany. *Imago Mundi, 65*(2), 253–267.

Dunn, S., & Wilkinson, S. M. (2013). Identifying critical components in infrastructure networks using network topology. *Journal of Infrastructure Systems, 19*(2), 157–165.

Ford, L. A., Crabtree, R. D., & Hubbell, A. (2009). Crossing borders in health communication research: Toward an ecological understanding of context, complexity, and consequences in community-based health education in the U.S.-Mexico borderlands. *Health Communication, 24*(7), 608–618.

Gelbman, A., & Timothy, D. J. (2011). Border complexity, tourism and international exclaves a case study. *Annals of Tourism Research, 38*(1), 110–131.

Hammons, T. J. (2010). Status of african and middle-east renewable energy projects: Infrastructure, developments and cross-border. *International Journal of Power and Energy Systems, 30*(2), 73–85.

Hillier, W. R. G., Turner, A., Yang, T., & Park, H. (2010). Metric and topo-geometric properties of urban street networks: Some convergencies, divergencies and new results. *Journal of Space Syntax, 1*(2), 258–279.

Jiang, B. (2007). A topological pattern of urban street networks: Universality and peculiarity. *Physica A: Statistical Mechanics and its Applications, 384*(2), 647–655.

Jiang, B., & Claramunt, C. (2004). Topological analysis of urban street networks. *Environment and Planning B: Planning and design, 31*(1), 151–162.

Jiang, B., & Liu, C. (2009). Street-based topological representations and analyses for predicting traffic flow in GIS. *International Journal of Geographical Information Science, 23*(9), 1119–1137.

Jiang, B., Yin, J., & Zhao, S. (2009). Characterizing the human mobility pattern in a large street network. *Physical Review E, 80*(2), 021136.

Jiang, B., Duan, Y., Lu, F., Yang, T., & Zhao, J. (2014). Topological structure of urban street networks from the perspective of degree correlations. *Environment and Planning B: Planning and Design, 41*(5), 813–828.

Khasnabish, B. (1989). Topological properties of Manhattan street networks. *Electronics Letters, 20*(25), 1388–1389.

Kirkley, A., Barbosa, H., Barthelemy, M., & Ghoshal, G. (2018). From the betweenness centrality in street networks to structural invariants in random planar graphs. *Nature Communications, 9*(1), 2501.

Kumar, A., & Rodrigues, J. M. (2009). Foreign currency-related translation complexities in cross-border healthcare applications. *Studies in Health Technology and Informatics, 150*, 96–100.

Lal, C. K. (2006). The complexities of border conflicts in South Asia. *South Asian Survey, 13*(2), 253–263.

Lama, M. P. (2000). Economic reforms and the energy sector in South Asia: Scope for cross-border power trade. *South Asian Survey, 7*(1), 3–23.

Lauritsen, K. (2012). Cultural complexity and border markers in Norwegian kindergartens. *Children and Society, 37*(5), 350–360.

Lee, B. H., & Jung, W. S. (2018). Analysis on the urban street network of Korea: Connections between topology and meta-information. *Physica A: Statistical Mechanics and its Applications, 497*, 15–25.

Li, J., Duenas-Osorio, L., Chen, C., & Shi, C. (2016). Connectivity reliability and topological controllability of infrastructure networks: A comparative assessment. *Reliability Engineering & System Safety, 156*, 24–33.

Lin, J., & Ban, Y. (2013). Complex network topology of transportation systems. *Transport Reviews, 33*(6), 658–685.

Lin, J., & Ban, Y. (2017). Comparative analysis on topological structures of urban street networks. *ISPRS International Journal of Geo-Information, 6*(10), 295.

Liu, W., & Song, Z. (2020). Review of studies on the resilience of urban critical infrastructure networks. *Reliability Engineering & System Safety, 193*, 106617.

Liu, Y.-Y., Slotine, J.-J., & Barabási, A.-L. (2011). Controllability of complex networks. *Nature, 473*, 167–173.

Masucci, A. P., Smith, D., Crooks, A., & Batty, M. (2009). Random planar graphs and the London street network. *The European Physical Journal B, 71*, 259–271.

Masucci, A. P., Stanilov, K., & Batty, M. (2014). Exploring the evolution of London's street network in the information space: A dual approach. *Physical Review E, 89*(1), 012805.

Opermanis, O., MacSharry, B., Aunins, A., & Sipkova, Z. (2012). Connectedness and connectivity of the Natura 2000 network of protected areas across country borders in the European Union. *Biological Conservation, 153*, 227–238.

Papadimitriou, F. (2015). 'Synoriology'–a science for the environment, peace, infrastructures and cross-border management. In *Environmental security of the European cross-border energy supply infrastructure* (pp. 187–191). Springer.

Papadimitriou, F. (2020). Modelling and visualization of landscape complexity with braid topology. In D. Edler, C. Jenal & O. Kühne (Eds), *Modern approaches to the visualization of landscapes* (pp. 79–101). Springer.

Rash, F. (2002). The German-romance language borders in Switzerland. *Journal of Multilingual and Multicultural Development, 23*(1–2), 112–136.

Saldarriaga, J., Zambrano, J., Herrán, J., & Iglesias-Rey, P. L. (2021). Layout selection for an optimal sewer network design based on land topography, streets network topology, and inflows. *Water, 13*(18), 2491.

Sarvasová, Z., Sálka, J., & Dobsinská, Z. (2012). Mechanism of cross-sectoral coordination between nature protection and forestry in the Natura 2000 formulation process in Slovakia. *Journal of Environmental Management, 127*, 65–72.

Schultz, P., Heitzig, J., & Kurths, J. (2014). A random growth model for power grids and other spatially embedded infrastructure networks. *The European Physical Journal Special Topics, 223*(12), 2593–2610.

Sharifi, A. (2019). Resilient urban forms: A review of literature on streets and street networks. *Building and Environment, 147*, 171–187.

Smith, S. (1999). Need for cross-border harmonization of energy taxes in an interdependent world. *International Journal of Global Energy Issues, 12*(7), 340–351.

Svendsen, N. K., & Wolthusen, S. D. (2007). Connectivity models of interdependency in mixed-type critical infrastructure networks. *Information Security Technical Report, 12*(1), 44–55.

Wang, J. (2015). Resilience of self-organised and top-down planned cities—A case study on London and Beijing street networks. *PLoS One, 10*(12), e0141736.
Wang, B., Gao, L., Gao, Y., & Deng, Y. (2013). Maintain the structural controllability under malicious attacks on directed networks. *EPL Europhysics Letters, 101*, 58003.
Watcharejyothin, M., & Shrestha, R. M. (2009). Effects of cross-border power trade between Laos and Thailand: Energy security and environmental implications. *Energy Policy, 37*(5), 1782–1792.
Zischg, J., Klinkhamer, C., Zhan, X., Rao, P. S. C., & Sitzenfrei, R. (2019). A century of topological coevolution of complex infrastructure networks in an alpine city. *Complexity*, Article ID 2096749.

Chapter 5
Geo-topology, Complexity and Resilience

Fools ignore complexity.
Pragmatists suffer it.
Some can avoid it.
Geniuses remove it.
(Alan Perlis, 1922–1990, "Epigrams on Programming")

5.1 Geo-topology and the Complexity of Landscape Change

One of the most important issues in landscape analysis is the assessment of landscape complexity (Wu & Hobbs, 2002; Papadimitriou, 2002; Honnay et al., 2003; Melles et al., 2003; Gabriel et al., 2005; Roschewitz et al., 2005; Concepción et al., 2008; Persson et al., 2010; Cormont et al., 2016; Nelson & Burchfield, 2021; Papadimitriou, 2012a, b, c, 2020d, f).

The complexity of changes in landscape boundaries can be measured by applying the Levenshtein distance Lev (or "edit distance"), that is the minimum number of operations which can transform one string of symbols to another string by either adding more symbols (a) to it, or deleting symbols from it (d) or replacing them by other ones (r):

$$\mathrm{Lev} = \min\{a + d + r\}.$$

For instance, the least costly in terms of operations way to convert the string of symbols x = AAABCCC to the string y = AAAABCA is by sliding the string y by one position to the left.

x AAABCCC
y AAAABCA

and then performing the following operations:

(i). Deletion of the first A of y:

x AAABCCC
y AAABCA

© The Author(s), under exclusive license to Springer Nature Switzerland AG 2023

F. Papadimitriou, *Geo-Topology*, GeoJournal Library 133,
https://doi.org/10.1007/978-3-031-48185-7_5

(ii). Replacement of the final A of y by a C:

x AAABCCC
y AAABCC

(iii). Addition of another C at the end of y.

x AAABCCC
y AAABCCC

Since three operations are required, the Levenshtein distance between the two strings is

$$\mathrm{Lev} = \min\{\mathrm{a} + \mathrm{d} + \mathrm{r}\} = 3$$

It has to be stressed that the algorithm for this measure seeks a minimum of operations and thus it is not a simple one-to-one comparison between the two strings of symbols (which is the Hamming distance). For instance, the Hamming distance counting all symbol-by-symbol differences between x and y is 4 (the Hamming distance is always greater than or equal to the Levenshtein distance):

x AAABCCC
y AAAABCA

(additionally, the calculation of the Hamming distance does not allow sliding one string with respect to another).

The Levenshtein distance has been used to check the regularity of contours (Abreu & Rico-Juan, 2011), to compare legal documents (Bernholz & Pytlik Zillig, 2011), to analyse longitudinal demographic data (Doran & Van Wamelen, 2010), dialectal separations between sibling languages (Palunčić et al., 2010) and in several technical applications, i.e. to measure travel flows of vehicle sequences (Takahashi & Izumi, 2006), to evaluate web security (Costa et al., 2011), in biometric applications such as iris-surveillance imagery (Uhl & Wild, 2010) etc. In spatial analysis, it was given a 2D expression (Papadimitriou, 2009) and in that form it can be used to measure the complexity of raster maps (Papadimitriou, 2020a, b, c, d, e). Besides its usefulness however, devising algorithms to efficiently calculate Levenshtein distances is a problem of its own in algorithmic complexity theory (Gali & Giancarlo, 1988; Goli & Petrovi, 1995; Pighizzini, 2001; Goli & Menicocci, 2002; Baake et al., 2006).

For the analysis of graphs of borders, the edit distance of two graphs G_1 and G_2 is defined (Bunke, 1997) as the shortest sequence of edit operations which can transform G_1 into G_2 (Riesen & Bunke, 2009; Blumenthal & Gamper, 2020; Blumenthal et al., 2020). The formula is essentially the same, with the only modification that it should account for both nodes and links:

$$\mathrm{Lev} = \min\{\mathrm{a}_{\mathrm{nodes}} + \mathrm{a}_{\mathrm{links}} + \mathrm{d}_{\mathrm{nodes}} + \mathrm{d}_{\mathrm{links}} + \mathrm{r}_{\mathrm{nodes}} + \mathrm{r}_{\mathrm{links}}\}.$$

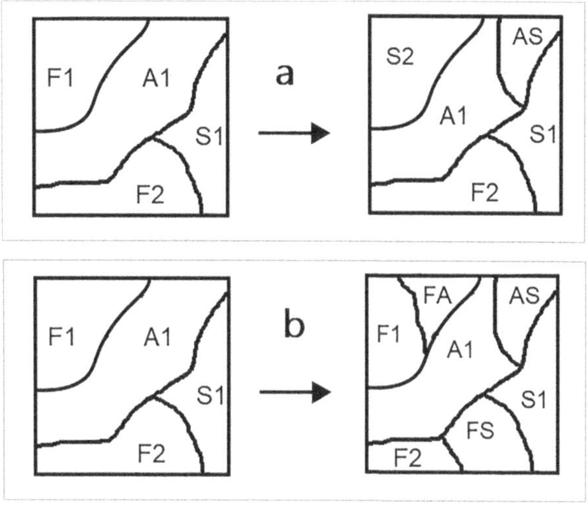

Fig. 5.1 Boundary changes of landscape L_1 from time 0 (left) to time 1 (case a). Codes stand for landscape types (F = forest, A = agriculture and S = shrubland) and number indices for patches (i.e. the forest patch F_1 is different from the forest patch F_2). The area that changed land cover from agriculture to shrubland is denoted by AS. In case b is shown a different landscape change scenario for landscape L_1

The graph edit distance is the most widely used measure of similarity between graphs (Sole-Ribalta & Serratosa, 2011) and has various applications in pattern recognition and machine vision (Gao et al., 2010), although its calculation is feasible for graphs of rather small size only (Robles-Kelly & Hancock, 2003; Riesen & Bunke, 2009). Consider, for instance, the boundary changes of an example landscape L_1 in the course of time (Fig. 5.1).

Applying the method of graph distances, the distance between the initial landscape and the final landscape is calculated by performing the following operations:

Replace node F_1 by S_2
Add link S_2-A_1
Add node AS
Add link A_1-S
Add link AS-S_1

and hence the graph edit distance is Lev=5 because this is the minimum number of operations (additions, deletions, replacements of nodes and links) required to convert the landscape L_1 from its state at time 0 to that at time 1.

Scenario (b) has the same landscape (L_1), but with different changes: part of the forested area F_1 has converted into agriculture (FA) and part of F_2 has converted into shrubland (SF). Consequently, the operations that correspond to these landscape changes are the following:

Insert node FA
Insert node AS
Insert node FS
Add link F_1-FA
Add link A_1-FA
Add link A_1-FS

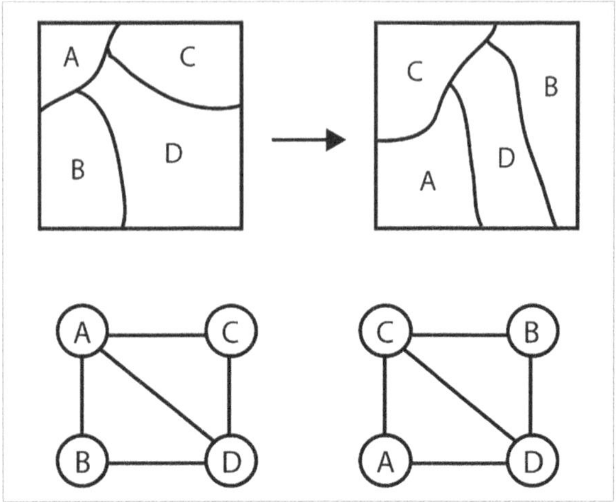

Fig. 5.2 Same connection topology but different land uses

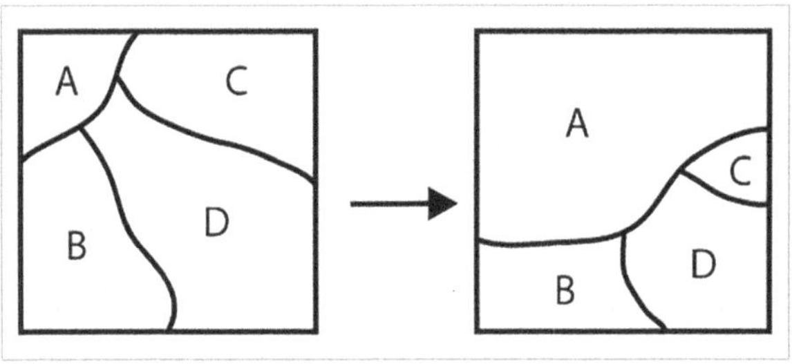

Fig. 5.3 Same types of land uses, but with different boundaries

Add link AS-S_1
Add link S_1-FS
Add link FS-F_2
Delete link F_2-S_1

and thus the graph edit distance in this case is Lev=10.

Given these, some situations need special attention, as they are often encountered in real geographical settings:

(i). Same connection topology but different land uses (Fig. 5.2).

This is a case that despite the two graphs being topologically isomorphic, their node labels are different. In order to transform the graph on the left to the graph on the right, the following operations are required:

Delete link AB
Insert link BC

and thus the graph edit distance is 2.

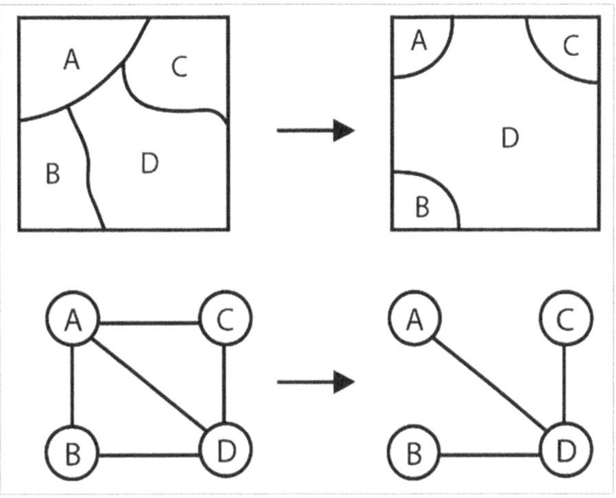

Fig. 5.4 Simplifying topology impacts graph distance

(ii). Same types of land uses, but with different boundaries (Fig. 5.3).

When only the topology of the boundaries changes, then the graph distance is zero. This reflects the fact that a mere change in area size without simultaneous change in connectivity does not affect the topology of the map, and therefore it does not affect the respective graphs either.

(iii). Simplifying topology affects graph distance (Fig. 5.4).

The disconnection between areas A and B, A and C results in a simpler graph and thus the operations required are:

Delete link AB
Delete link AC

(iv). Bivalent links (Fig. 5.5).

In case there are two different patches of land use C (the upper C patch borders both A and B, while the lower part borders both B and D), then there is a bivalent link of C.

The additions or deletions of bivalent links in calculations of graph distance must be carried out by separate operations. Transforming the map on the left to the map on the right involves the following four operations:

Add AC link
Delete AD link
Add CB link
Add CB link

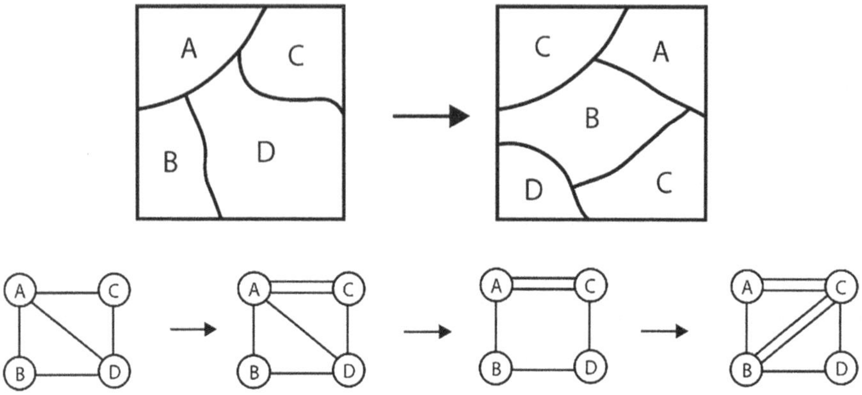

Fig. 5.5 Bivalent links: an example transformation

But if the lower right C patch is considered to be a different land use type (E instead of C), then the number of graph operations increases up to 7 (Fig. 5.6):

Delete AD link
Add BC link
Delete CD link
Add node E
Add link AE
Add link EB
Add link ED

5.2 Resilience of Landscape Boundaries

Topological changes in landscape boundaries can be indicators of both landscape complexity and landscape resilience. Most commonly, in studying the resilience of landscape networks, we are interested in the impact of the removal of one or more nodes on the stability and resilience of the remaining network. The topology of a network can be an indirect indicator of disturbance. An external disturbance (i.e. infection) spreads easier (and affects network nodes easier) if they are more easily accessible, while sparse networks can be more protected (Wan et al., 2021). Besides, problems of resilience may also be examined by adopting a complex systems approach (Berkes et al., 2003; Selkirk et al., 2018; Asllani & Carletti, 2018; Pettersen Gould, 2019), as in studies of resilience of urban systems (McPhearson et al., 2016, 2021; Sauter et al., 2021). While it can be accepted that high network connectivity facilitates recovery after disturbance, it is most likely a question of "how much", because highly connected systems may lead to synchronized behaviors and thus become less resilient (Van Nes & Scheffer,

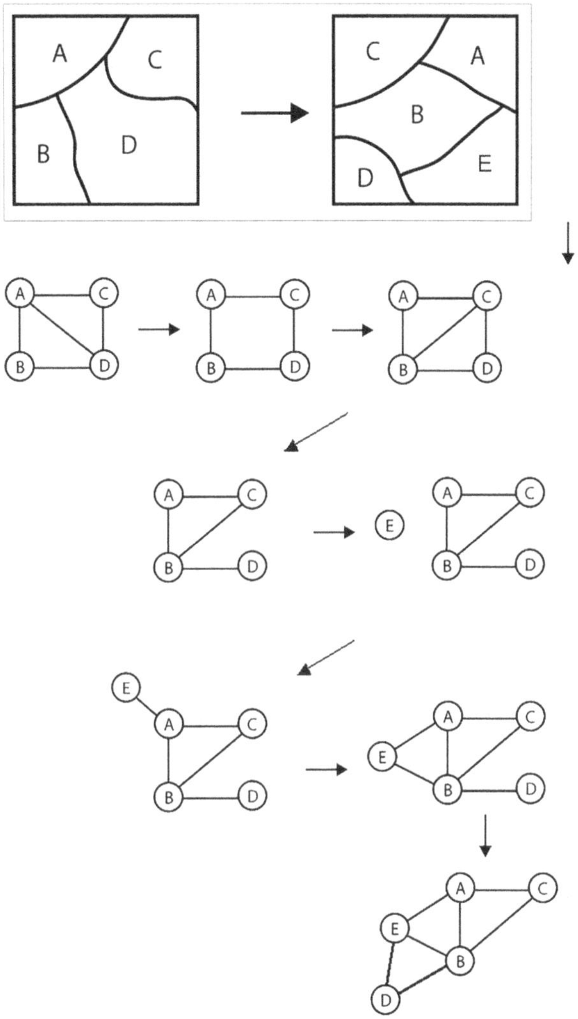

Fig. 5.6 The calculation process with the introduction of the patch E

2005; Gelcich et al., 2006; Satake et al., 2007; Adger et al., 2009; Biggs et al., 2011, 2012). Landscape resilience (Holling, 1973; Peterson et al., 1998; Scheffer et al., 2015; Meyer, 2016; Van de Leemput et al., 2018; Assumma et al., 2018; Bengtsson et al., 2021; Papadimitriou, 2023a, b) is examined at various scales and contexts, i.e. socio-ecological (Cumming et al., 2013), cultural (Henning et al., 2021), in the context of land use planning (White & O'Hare, 2014), or forest resilience (Arianoutsou et al., 2011; Verbesselt et al., 2016; Lucash et al., 2017). Modelling the adjacencies among patches as networks, consider for instance, the maps of land use change of an area west of Sounio (60 km southeast of Athens, Greece) for which the maps of land use changes from 1967 through 1988 to 2018 (Fig. 5.7) have been

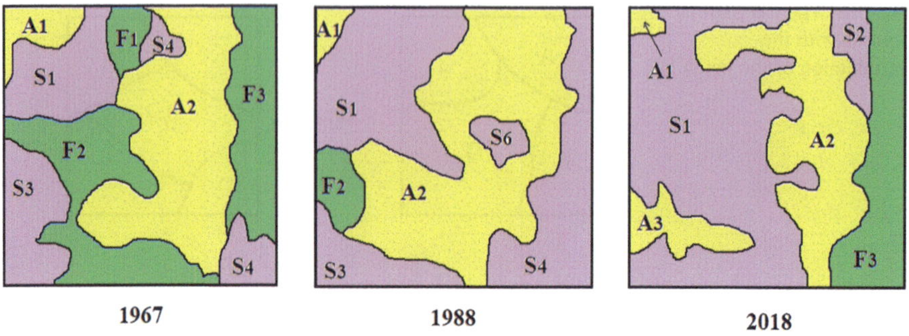

Fig. 5.7 Maps of land use change of an area close to Sounio (Athens, Greece): three land use types (F = forest, A = agriculture, S = shrublands) with numbered patches

derived from various sources: (aerial photographs, satellite imagery and field mapping).

The topological analysis of boundaries and patches can provide an effective assessment of the resilience of each one of these three land use types. The first step is to build network models reflecting the structure of landscape boundaries at each one of the three time instants (Fig. 5.8).

The mean degree of each land use type and its changes through time can be suggestive of landscape resilience: while the mean degree per patch decreased in forests and agriculture, it appears resilient in the case of shrublands, which increased their degree per patch from 1988 to 2018 (Fig. 5.9).

A question then raises as to the geographic and ecological explanation of the differences in resilience among these three land use types.

The typical forest species of the Sounio area is the *Pinus halepensis*, which is often encountered along with Mediterranean cypress (*Cypressus sempervirens*) or stone pines (*Pinus pinea* L.). Other characteristic forest species of the bioclimatic zone of Sounio are the pine trees (*Pinus halepensis* mainly and *Pinus brutia*). Some of the forest areas however, are due to reforestation.

The Mediterranean olive (*Olea europaea*) dominates the agricultural lands, along with vineyards. Olive trees are also found as components of natural forests or, quite often, amidst shrublands also. This is because *Olea* is a very strong, drought-resistant species which survives well, has effectively colonized the lands of Sounio, withstanding land degradation successfully.

In ecotones of *F-S* or *A-S* patches, other species that naturally grow in East Attica are the *Juniperus oxycedrus, Platanus sp., Arbutus sp., Ficus carica, Eucalyptus globulus*. The very dynamic shrublands (the *S* landscape type) of Sounio contain two types of vegetation communities: the "maquis" and the "phrygana". The maquis (*Quercus ilex, Pistacia lentiscus* etc) are a dense and resistant scrub formation, comprising also species of myrtle, lavender, rosemary, marjoram, thyme etc.

The degraded shrubland consists of phrygana vegetation with main representative the *Sarcopoterium spinosum, Phlomis fructicosa* and the *Genista acantholada*, mainly growing on the sunny sides of hilly and downslope regions. The phrygana

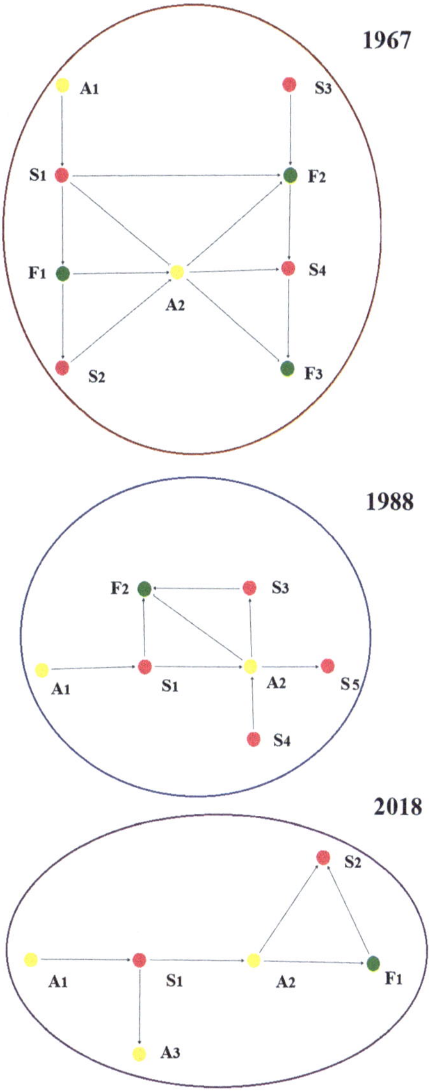

Fig. 5.8 Network models of the patches for each one of the years of observation of the Sounio landscape (1967, 1988 and 2018)

are not only suited to the pedological or climatic conditions, but are also a remarkably resilient landscape type after multiple disturbances.

The human impact on the vegetation in Sounio (the exogenous disturbance to this landscape) can be traced from within the maquis-to-phrygana land degradation sequences, which are typical (with local variations) for Mediterranean landscapes. In a way, the resilience of shrublands, as assessed from the topology of boundaries,

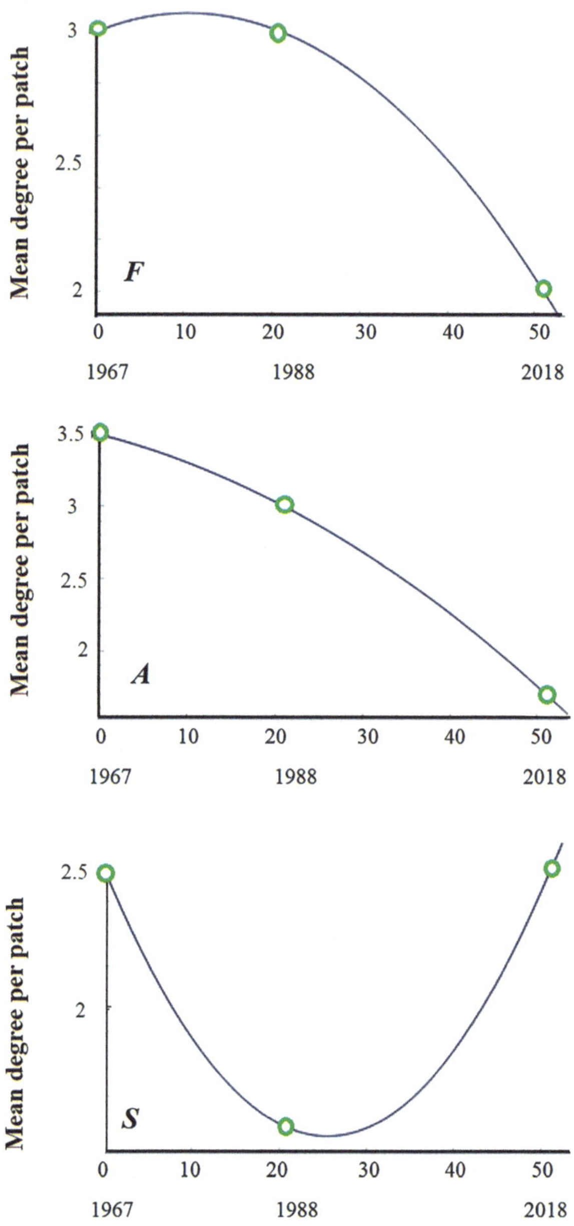

Fig. 5.9 Changes in the mean degree per land use type in Sounio from 1967 to 2018: decrease in forests and agriculture but increase in shrublands

reflects the differences that the species of shublands have (in terms of ecophysiological properties, adaptability and ecological functions) compared to agriculture and forests.

References

Abreu, J., & Rico-Juan, J. R. (2011). Characterization of contour regularities based on the Levenshtein edit distance. *Pattern Recognition Letters, 32*(10), 1421–1427.

Adger, W. N., Eakin, H., & Winkels, A. (2009). Nested and teleconnected vulnerabilities to environmental change. *Frontiers in Ecology and the Environment, 7*, 150–157.

Arianoutsou, M., Koukoulas, S., & Kazanis, D. (2011). Evaluating post-fire forest resilience using GIS and multi-criteria analysis: An example from Cape Sounion National Park, Greece. *Environmental Management, 47*(3), 384–397.

Asllani, M., & Carletti, T. (2018). Topological resilience in non-normal networked systems. *Physical Review E, 97*(4), 042302.

Assumma, V., Bottero, M., Monaco, R., & Soares, A. J. (2018). An integrated evaluation model for shaping future resilient scenarios in multi-pole territorial systems. *Environmental and Territorial Modelling for Planning and Design, 4*, 17–24.

Baake, M., Grimm, U., & Giegerich, R. (2006). Surprises in approximating Levenshtein distances. *Journal of Theoretical Biology, 243*(2), 279–282.

Bengtsson, J., Angelstam, P., Elmqvist, T., Emanuelsson, U., Folke, C., Ihse, M., et al. (2021). Reserves, resilience and dynamic landscapes 20 years later. *Ambio, 50*(5), 962–966.

Berkes, F., Colding, J., & Folke, C. (2003). *Navigating social–ecological systems: Building resilience for complexity and change*. Cambridge University Press.

Bernholz, C. D., & Pytlik Zillig, B. L. (2011). Comparing nearly identical treaty texts: A note on the treaty of Fort Laramie with Sioux, etc., 1851 and Levenshtein's edit distance metric. *Literary and linguistic computing, 26*(1), 5–16.

Biggs, D., Biggs, R., Dakos, V., Scholes, R. J., & Schoon, M. L. (2011). Are we entering an era of concatenated global crises? *Ecology and Society, 14*, 32.

Biggs, R., Schlüter, M., Biggs, D., Bohensky, E. L., Burn Silver, S., Cundill, G., et al. (2012). Toward principles for enhancing the resilience of ecosystem services. *Annual Review of Environment and Resources, 37*, 421–448.

Blumenthal, D. B., & Gamper, J. (2020). On the exact computation of the graph edit distance. *Pattern Recognition Letters, 134*, 46–57.

Blumenthal, D. B., Boria, N., Gamper, J., Bougleux, S., & Brun, L. (2020). Comparing heuristics for graph edit distance computation. *The VLDB Journal, 29*(1), 419–458.

Bunke, H. (1997). On a relation between graph edit distance and maximum common subgraph. *Pattern Recognition Letters, 18*(8), 689–694.

Concepci6n, E. D., Díaz, M., & Baquero, R. A. (2008). Effects of landscape complexity on the ecological effectiveness of Agri-environment schemes. *Landscape Ecology, 23*, 135–148.

Cormont, A., Siepel, H., Clement, J., Melman, T. C., Wallis DeVries, M. F., Van Turnhout, C. A. M., et al. (2016). Landscape complexity and farmland biodiversity: Evaluating the CAP target on natural elements. *Journal for Nature Conservation, 30*, 19–26.

Costa, D., Pirmez, L., Carmo, L. F. R. C., & Bacellar, L. F. H. (2011). Enhancing Levenshtein distance algorithm for assessing behavioral trust. *Computer Systems Science and Engineering, 26*(2), 77–85.

Cumming, G. S., Olsson, P., Chapin, F. S., & Holling, C. S. (2013). Resilience, experimentation, and scale mismatches in social-ecological landscapes. *Landscape Ecology, 28*(6), 1139–1150.

Doran, H. C., & Van Wamelen, P. B. (2010). Application of the Levenshtein distance metric for the construction of longitudinal data files. *Educational Measurement: Issues and Practice, 29*(2), 13–23.

Gabriel, D., Thies, C., & Tscharntke, T. (2005). Local diversity of arable weeds increases with landscape complexity. *Perspectives in Plant Ecology, Evolution and Systematics, 7*(2), 85–93.

Gali, Z., & Giancarlo, R. (1988). Data structures and algorithms for approximate string matching. *Journal of Complexity, 4*(1), 33–72.

Gao, X., Xiao, B., Tao, D., & Li, X. (2010). A survey of graph edit distance. *Pattern Analysis and Applications, 13*(1), 113–129.

Gelcich, S., Edwards-Jones, G., Kaiser, M. J., & Castilla, J. C. (2006). Co-management policy can reduce resilience in traditionally managed marine ecosystems. *Ecosystems, 9*, 951–966.
Goli, J. D., & Menicocci, R. (2002). Computation of edit probabilities and edit distances for the A5-type keystream generator. *Journal of Complexity, 18*(1), 356–374.
Goli, J. D., & Petrovi, S. V. (1995). Constrained many-to-one string editing with memory. *Information Sciences, 86*(1–3), 61–76.
Henning, M., Herrmann, P., Zimmermann, T., Meier, M., Pietsch, M., & Schmidt, C. (2021). A scenario and monitoring based planning approach to strengthen the resilience of the cultural landscape. *Journal of Digital Landscape Architecture, 6-2021*, 123–132.
Holling, C. S. (1973). Resilience and stability of ecological systems. *Annual Review of Ecology and Systematics, 4*(1), 1–23.
Honnay, O., Piessens, K., Van Landuyt, W., Hermy, M., & Gulinck, H. (2003). Satellite based land use and landscape complexity indices as predictors for regional plant species diversity. *Landscape and Urban Planning, 63*(4), 241–250.
Lucash, M. S., Scheller, R. M., Gustafson, E., & Sturtevant, B. (2017). Spatial resilience of forested landscapes under climate change and management. *Landscape Ecology, 32*(5), 953–969.
McPhearson, T., Iwaniec, D. M., & Bai, X. (2016). Positive visions for guiding urban transformations toward sustainable futures. *Current Opinion in Environmental Sustainability, 22*, 33–40.
McPhearson, T., Iwaniec, D. M., Hamstead, Z. A., Berbés-Blázquez, M., Cook, E. M., Muñoz-Erickson, T. A., et al. (2021). A vision for resilient urban futures. In Z. A. Hamstead, D. M. Iwaniec, T. McPhearson, M. Berbés-Blázquez, E. M. Cook, & T. A. Muñoz-Erickson (Eds.), *Resilient urban futures* (pp. 173–186). Springer International Publishing.
Melles, S., Glenn, S., & Martin, K. (2003). Urban bird diversity and landscape complexity: Species–environment associations along a multiscale habitat gradient. *Conservation Ecology, 7*(1), 5.
Meyer, K. (2016). A mathematical review of resilience in ecology. *Natural Resources Modeling, 29*, 339–352.
Nelson, K. S., & Burchfield, E. K. (2021). Landscape complexity and US crop production. *Nature Food, 2*(5), 330–338.
Palunčić, F., Ferreira, H. C., Swart, T. G., & Clarke, W. A. (2010). Modelling distances between genetically related languages using an extended weighted Levenshtein distance. *Southern African Linguistics and Applied Language Studies, 27*(4), 381–389.
Papadimitriou, F. (2002). Modelling indicators and indices of landscape complexity: An approach using GIS. *Ecological Indicators, 2*(1–2), 17–25.
Papadimitriou, F. (2009). Modelling spatial landscape complexity using the Levenshtein algorithm. *Ecological Informatics, 4*(1), 48–55.
Papadimitriou, F. (2012a). Artificial intelligence in modelling the complexity of Mediterranean landscape transformations. *Computers and Electronics in Agricutlure, 81*, 87–96.
Papadimitriou, F. (2012b). Modelling landscape complexity for land use Management in Rio de Janeiro, Brazil. *Land Use Policy, 29*(4), 855–861.
Papadimitriou, F. (2012c). The algorithmic complexity of landscapes. *Landscape Research, 37*(5), 599–611.
Papadimitriou, F. (2020a). *Spatial complexity: Theory, mathematical methods and applications*. Springer.
Papadimitriou, F. (2020b). The algorithmic basis of spatial complexity. In *Spatial complexity: Theory, mathematical methods and applications* (pp. 81–99). Springer.
Papadimitriou, F. (2020c). The spatial complexity of 3×3 binary maps. In *Spatial complexity: Theory, mathematical methods and applications* (pp. 163–178). Springer.
Papadimitriou, F. (2020d). Spatial complexity in nature, science and technology. In *Spatial complexity: Theory, mathematical methods and applications* (pp. 19–35). Springer.
Papadimitriou, F. (2020e). The topological basis of spatial complexity. In *Spatial complexity. Theory, mathematical methods and applications* (pp. 63–79). Springer.

Papadimitriou, F. (2020f). Modelling and visualization of landscape complexity with braid topology. In *Modern approaches to the visualization of landscapes* (pp. 79–101). Springer.
Papadimitriou, F. (2023a). Modelling landscape resilience. In *Modelling landscape dynamics. Determinism, stochasticity, complexity*. Springer.
Papadimitriou, F. (2023b). Landscape stability, instability and civilization collapse. In *Modelling landscape dynamics. Determinism, stochasticity, complexity*. Springer.
Persson, A. S., Olsson, O., Rundlöf, M., & Smith, H. G. (2010). Land use intensity and landscape complexity—Analysis of landscape characteristics in an agricultural region in southern Sweden. *Agriculture, Ecosystems & Environment, 136*(1–2), 169–176.
Peterson, G., Allen, C. R., & Holling, C. S. (1998). Ecological resilience, biodiversity, and scale. *Ecosystems, 1*(1), 6–18.
Pettersen Gould, K. (2019). Precursor resilience in practice–an organizational response to weak signals. In *Exploring Resilience: A Scientific Journey from Practice to Theory* (pp. 51–58). Springer.
Pighizzini, G. (2001). How hard is computing the edit distance? *Information and Computation, 165*, 1–13.
Riesen, K., & Bunke, H. (2009). Approximate graph edit distance computation by means of bipartite graph matching. *Image and Vision Computing, 27*(7), 950–959.
Robles-Kelly, A., & Hancock, E. R. (2003). Edit distance from graph spectra. In *Proceedings of the IEEE International Conference on Computer Vision 1* (pp. 234–241). Nice.
Roschewitz, I., Gabriel, D., Tscharntke, T., & Thies, C. (2005). The effects of landscape complexity on arable weed species diversity in organic and conventional farming. *Journal of Applied Ecology, 42*(5), 873–882.
Satake, A., Leslie, H. M., Iwasa, Y., & Levin, S. A. (2007). Coupled ecological–social dynamics in a forested landscape: Spatial interactions and information flow. *Journal of Theoretical Biology, 246*(4), 695–707.
Sauter, D., Randhawa, J., Tomateo, C., & McPhearson, T. (2021). Visualizing urban social–ecological–technological systems. In *Resilient Urban Futures* (pp. 145–157). Springer.
Scheffer, M., Carpenter, S. R., Dakos, V., & van Nes, E. H. (2015). Generic indicators of ecological resilience: Inferring the chance of a critical transition. *Annual Review of Ecology, Evolution, and Systematics, 46*(1), 145–167.
Selkirk, K., Selin, C., & Felt, U. (2018). A festival of futures: Recognizing and reckoning temporal complexity in foresight. In *Handbook of anticipation: Theoretical and applied aspects of the use of future in decision making* (pp. 1–23). Springer.
Sole-Ribalta, A., & Serratosa, F. (2011). Exploration of the labelling space given graph edit distance costs. *Lecture Notes in Computer Science, 6658*, 164–174.
Takahashi, S., & Izumi, T. (2006). Travel time measurement by vehicle sequence matching method-evaluation method of vehicle sequence using Levenshtein distance. In *2006 SICE-ICASE international joint conference* (pp. 1625–1629). IEEE.
Uhl, A., & Wild, P. (2010). Enhancing iris matching using levenshtein distance with alignment constraints. In *Advances in visual computing: 6th international symposium, ISVC 2010*. November 29–December 1, 2010. Proceedings, Part I 6 (pp. 469–478). Springer.
Van de Leemput, I. A., Dakos, V., Scheffer, M., & van Nes, E. H. (2018). Slow recovery from local disturbances as an indicator for loss of ecosystem resilience. *Ecosystems, 21*(1), 141–152.
Van Nes, E. H., & Scheffer, M. (2005). Implications of spatial heterogeneity for catastrophic regime shifts in ecosystems. *Ecology, 86*(7), 1797–1807.
Verbesselt, J., Umlauf, N., Hirota, M., Holmgren, M., Van Nes, E. H., Herold, M., et al. (2016). Remotely sensed resilience of tropical forests. *Nature Climate Change, 6*(11), 1028–1031.
Wan, Z., Mahajan, Y., Kang, B. W., Moore, T. J., & Cho, J. H. (2021). A survey on centrality metrics and their network resilience analysis. *IEEE Access, 9*, 104773–104819.

White, I., & O'Hare, P. (2014). From rhetoric to reality: Which resilience, why resilience, and whose resilience in spatial planning? *Environment and Planning C: Government and Policy, 32*(5), 934–950.

Wu, J., & Hobbs, R. (2002). Key issues and research priorities in landscape ecology: An idiosyncratic synthesis. *Landscape Ecology, 17*, 355–365.

Part III
Topology in Geo-Visualization

Chapter 6
Geo-topological Visualization of Landscapes and Landforms

Everything happens by following numbers
"Πάντα κατ' ἀριθμόν γίγνονται"
(Pythagoras, 580-496 b.C.)

6.1 Topological Surfaces

One of the most interesting features of topology is its focus on "topological surfaces" of which the *Möbius band* is probably the most widely known: a non-orientable (one-faced) surface in which it is impossible to distinguish an "outside" from an "inside" and can be created by using a set of parametric equations (Fig. 6.1a):

$$x(u,v) = \left[R + \frac{v}{2}\cos\left(\frac{u}{2}\right)\right]\cos u$$
$$y(u,v) = \left[R + \frac{v}{2}\cos\left(\frac{u}{2}\right)\right]\sin u$$
$$z(u,v) = v\sin\frac{u}{2}$$

Another important topological surface is the *torus*; a donut-shaped object with one hole (Fig. 6.1b), which can be made by adding a handle to a sphere and can be created by means of a set of parametric equations of the type:

$$x = (c + a\cos v)\cos u$$
$$y = (c + a\cos v)\sin u$$
$$z = a\sin v$$

with u and v taking values in the interval $[0, 2\pi)$.

Objects of different genus (number of holes or cavities in them) are not *homeomorphic* to one another. Thus, objects without holes are homeomorphic, which implies that they can be transformed to one another without tearing apart or punching them (Fig. 6.2). Equivalently, objects of different genus can not be homeomorphic.

© The Author(s), under exclusive license to Springer Nature Switzerland AG 2023
F. Papadimitriou, *Geo-Topology*, GeoJournal Library 133,
https://doi.org/10.1007/978-3-031-48185-7_6

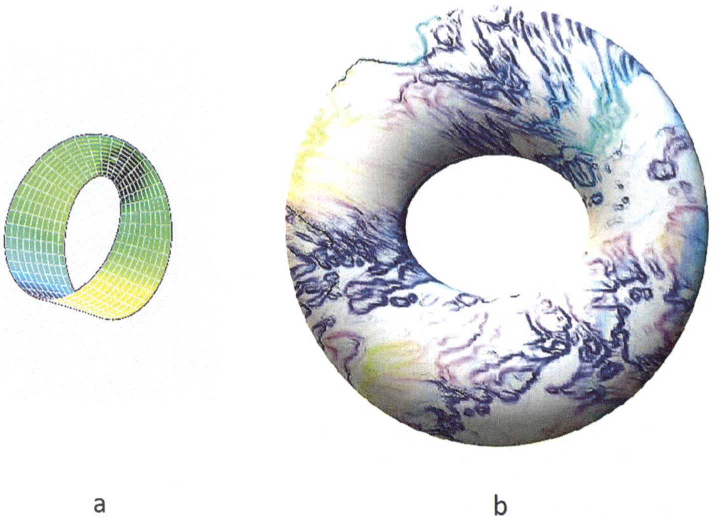

Fig. 6.1 *A* Möbius band (**a**) and a torus (**b**)

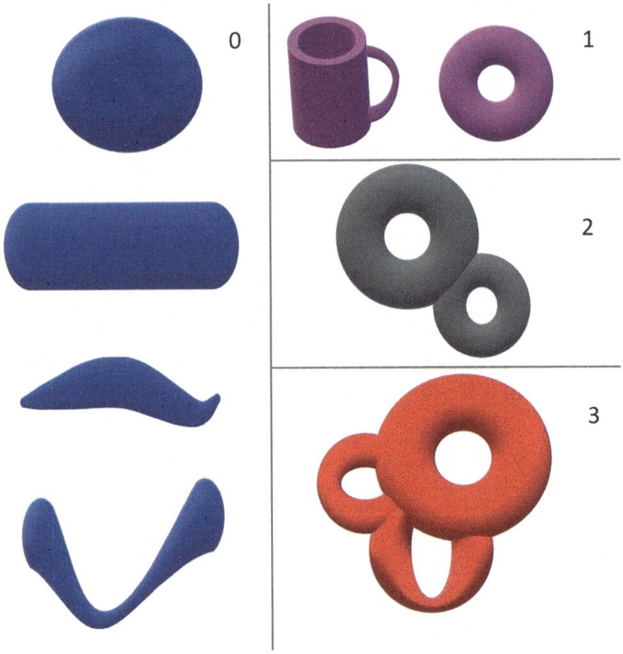

Fig. 6.2 Various 3D shapes that are homeomorphic to one another, depending on their genus: only those of the same genus are homeomorphic to one another (i.e. the cup is homeomorphic to the torus, but not to the double-torus)

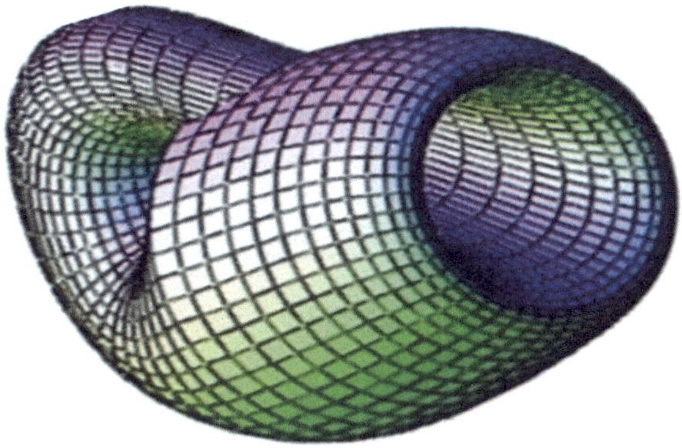

Fig. 6.3 The Klein bottle is a surface immersed in 3D space

A closed surface can be considered as a topologically compact and connected surface to which the fundamental "Dehn's classification theorem" applies: "every compact, connected surface is topologically equivalent to a sphere, or to a connected sum of tori, or to a connected sum of projective planes". All these classes of surfaces are non-homeomorphic to one another; homeomorphic surfaces have homeomorphic boundaries, so the cylinder and the Möbius strip are not homeomorphic to one another.

An "*embedding*" can place a surface or a spatial object in a space in such a way that topological properties (i.e. its connectivity) of that surface or object are preserved. A simple graph can be embedded onto a sphere without crossings, but more complex graphs require a hole for their embedding in order to maintain their connectivity and to avoid self-intersections. In such cases, a sphere may not be appropriate for their embedding and such surfaces may need a torus, or a double torus, or even other higher genus spatial domains to be embedded in.

If a surface is "*immersed*" in a n-dimensional space, it can not be embedded without self-intersection in that space. A *Klein bottle* is such a topological surface in 3D (Fig. 6.3), which can be described by a set of three parametric equations:

$$x = \left(r + \cos\frac{\theta}{2}\sin u - \sin\frac{\theta}{2}\sin 2u\right)\cos\theta$$

$$y = \left(r + \cos\frac{\theta}{2}\sin u - \sin\frac{\theta}{2}\sin 2u\right)\sin\theta$$

$$z = \sin\frac{\theta}{2}\sin u + \cos\frac{\theta}{2}\sin 2u$$

$$0 \le \theta \le 2\pi,$$

$$0 \le u < 2\pi$$

Although it can not be embedded in 3D space, it can nevertheless be embedded in a 4D space (without self-intersections), in which case it is describable by a different set of parametric equations:

$$x = r_1\left(\cos\frac{\theta}{2}\cos u - \sin\frac{\theta}{2}\sin 2u\right)$$
$$y = r_1\left(\sin\frac{\theta}{2}\cos u - \cos\frac{\theta}{2}\sin 2u\right)$$
$$z = r_2\cos\theta(1 + \sin u)$$
$$w = r_2\sin\theta(1 + \sin u)$$

6.2 Critical Points and Fixed Point Theorems

The vector field of the function $F(x,y)$ is:

$$\mathrm{V} = \left(\frac{\partial F}{\partial \mathrm{y}}, -\frac{\partial F}{\partial x}\right)$$

The integral paths of V are the contour lines of $F(x,y)$.

Then, the field is defined by the "gradient" of F (Fig. 6.4):

$$\nabla \mathrm{F}(x, y) = \left(\frac{\partial \mathrm{F}}{\partial \mathrm{x}}, \frac{\partial \mathrm{F}}{\partial y}\right)$$

and its integral curves are the lines of steepest ascent (in physical geography, we know such lines as equidistant, isobaric isolines etc).

For an equation $F(x,y)$, the *contour lines* of constant height are the solutions of the equation $F(x,y) =$ constant. If F is differentiable,

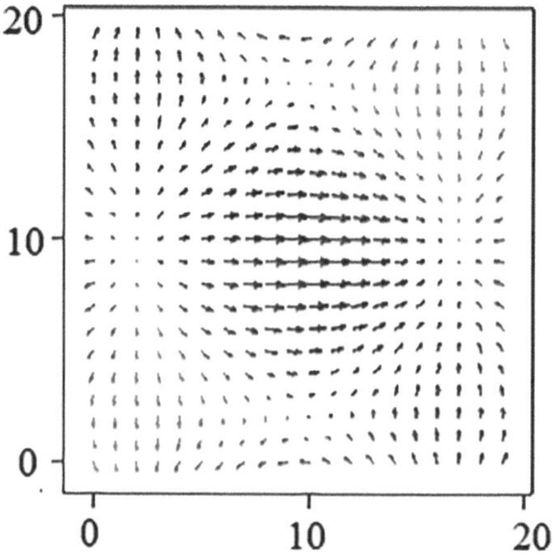

Fig. 6.4 The gradient field of the trigonometric function $f(x, y) = 2\sin x + \cos(y/2)$

then

$$\frac{\partial F}{\partial x}dx + \frac{\partial F}{\partial y}dy = 0$$

and the contours satisfy the equation

$$\frac{dy}{dx} = -\left(\frac{\frac{\partial F}{\partial x}}{\frac{\partial F}{\partial y}}\right).$$

Hence, the contour lines are the integral paths of the vector field

$$\mathrm{V}(x,y) = \left(\frac{\partial F}{\partial y}, -\frac{\partial F}{\partial x}\right)$$

The critical points of the function V are either centers (peaks or pits) or saddle points (the critical points are where $F = 0$) and the function

$$\nabla\mathrm{V}(x,y) = \left(\frac{\partial F}{\partial x}, \frac{\partial F}{\partial y}\right)$$

is the "gradient" of F.

The integral curves of ∇F are the lines showing the steepest ascent on the anaglyph (the curves of V meet those of ∇F at right angles).

Now let V be a continuous vector field

$$\mathrm{V}(x,y) = (F(x,y), G(x,y)),$$

defined by a system of differential equations:

$$\frac{dx}{dt} = F(x,y)$$
$$\frac{dy}{dt} = G(x,y)$$

and let D be a cell and γ its boundary.

The curves of dynamical system in the phase space may be winding (turning) around some points. Then, the winding number of a closed curve around a point is an integer number that reflects the number of times that the curve coils counterclockwise (or negative number if counterclockwise) around that point. If the vector field V is non-zero on the boundary, then if P_1, P_2, ..., P_n are *critical points* of V, the "*Poincaré index theorem*" assures that the "winding number W" is:

$$\mathrm{W}(\gamma) = \mathrm{I}(\mathrm{P}_1) + \mathrm{I}(\mathrm{P}_2) + \ldots + \mathrm{I}(\mathrm{P}_n).$$

This theorem connects the behavior of the vector field inside the cell with its behavior on the boundary, stating that if V is a continuous tangent vector field on a compact and connected orientable surface, then the sum of the indexes of the critical points of V equals the Euler characteristic of the surface. Further, the "*Poincaré-Brouwer theorem*" asserts that every continuous tangent vector field on a sphere has a critical point. This is why it is also called the "hairy ball theorem" (it is impossible to comb a hairy ball without leaving a point uncombed).

If the vector of the partial derivatives (the gradient) of a function F is zero at some point, then that point is "critical". This depends on whether the determinant of the Hessian matrix at that point is zero or not. If F is a function from R^n to R and all its second order partial derivatives exist, then its Hessian matrix is a nxn matrix defined as:

$$\mathbf{H}_F = = \begin{bmatrix} \frac{\partial^2 F}{\partial x_1^2} & \frac{\partial^2 F}{\partial x_1 \partial x_2} & \cdots & \frac{\partial^2 F}{\partial x_1 \partial x_n} \\ \frac{\partial^2 F}{\partial x_2 \partial x_1} & \frac{\partial^2 F}{\partial x_2^2} & \cdots & \frac{\partial^2 F}{\partial x_2 \partial x_n} \\ \cdots & \cdots & \cdots & \cdots \\ \frac{\partial^2 F}{\partial x_n \partial x_1} & \frac{\partial^2 F}{\partial x_n \partial x_2} & \cdots & \frac{\partial^2 F}{\partial x_n^2} \end{bmatrix}$$

The Hessian is the transpose of the Jacobian of F:

$$\mathbf{H}_F = \mathbf{J}(\nabla F)^{\mathrm{T}}$$

The classification of critical points depends on the determinant of the Hessian. If the determinant of the Hessian is zero, then the point is a "degenerate critical point" but if it is non-degenerate, then it is a "Morse critical point" of F. The number of negative eigenvalues of the determinant of the Hessian at some point of F determines the "index" of that point. The Hessian and the Morse critical points are the conceptual premises for constructing a "*Reeb graph*", which is the quotient space defined by the equivalence relation that identifies the points belonging to the same connected components of the level sets of F (Adelson-Velskii & Kronrod, 1945; Reeb, 1946; Biasotti et al., 2008). Each node of the Reeb graph corresponds to a critical point and each arc of the connected component of the manifold corresponds to two critical levels of the height function.

A *topological invariant* of a surface is a characteristic that remains invariant over all continuous transformations of that surface. One such invariant is the Euler characteristic (χ) and for every graph on the surface of the sphere: $\chi = v\text{-}e + f$, where v = vertices, e = edges, f = faces of the geometric object. Thus, $\chi = 2$ for the sphere, $\chi = 0$ for the torus and the Klein bottle, $\chi = 2\text{-}2n$ for a sphere with n-handles and $\chi = 2\text{-}n$ for a sphere with n-Möbius strips.

Closely related to the Poincaré-Brouwer theorem is "*Brouwer's Fixed Point Theorem*": imagine that a sheet of paper (e.g. a map) stretched on a table and that both the table and the sheet of paper have the same size and dimensions (length and width). If we lift the sheet of paper, fold it, bend it and crumble it in any way and throw it again on the table, then, following this theorem, there is one point from the sheet of paper which still has a one-to-one direct correspondence with one point on the table beneath, as it had before it was folded. Following this remarkable theorem, for each n the n-ball B^n has the fixed point property. *For the plane, the theorem proves that every continuous function f mapping the disk to itself, has a fixed point.* Formally, let f be a function mapping a topological space X to itself. If $f(x)$ for some x in X, then x maps to itself under f and is "a fixed point" of f. The fixed point property is a topological property, therefore if two spaces X and Y are homeomorphic, then X has the fixed point property if and only if Y does.

There are several physical applications of Brouwer's theorem. A typical one, from meteorology, is that there is always one point on the surface of the earth where the wind blows vertically and hence there is a cyclone at any moment in the atmosphere. So when comparing two maps of a landscape for times t_1 and t_2, at least one point of the landscape has remained unchanged during the landscape transformation process. That fixed point exists either if the landscape transformation involved land cover/land use changes or changes in the landscape's geomorphic systems, or both.

6.3 Euler Numbers and Reeb Graphs

The sum of the indexes of the critical points is equivalent to the Euler characteristic of the surface. If G is the height function on a clearly defined geographical space with boundary (such as an island) and V a vector field over that geographical space, then the winding number of V along the boundary of that geographical space (or the shoreline, if it is an island) is expected to be 1. Considering that the critical points of the vector field are peaks (P), bottoms (B), and saddle points (S), we have:

$$P + B - S = 1.$$

Further, if this space has internal boundaries also (i.e lakes, L), then

$$P + B - S = 1 - L.$$

More precisely, the Euler number of a 3D surface S is

$$\chi(S) = \text{Peaks} + \text{Pits} - \text{Passes}.$$

The value of $\chi(S)$ is 2 for the sphere (and surfaces homeomorphic to it), 0 for the torus and 2-2g for a connected sum of g-tori (or a sphere with g handles). Euler's number assumes different values for different surfaces: i.e. it is 0 for the Möbius

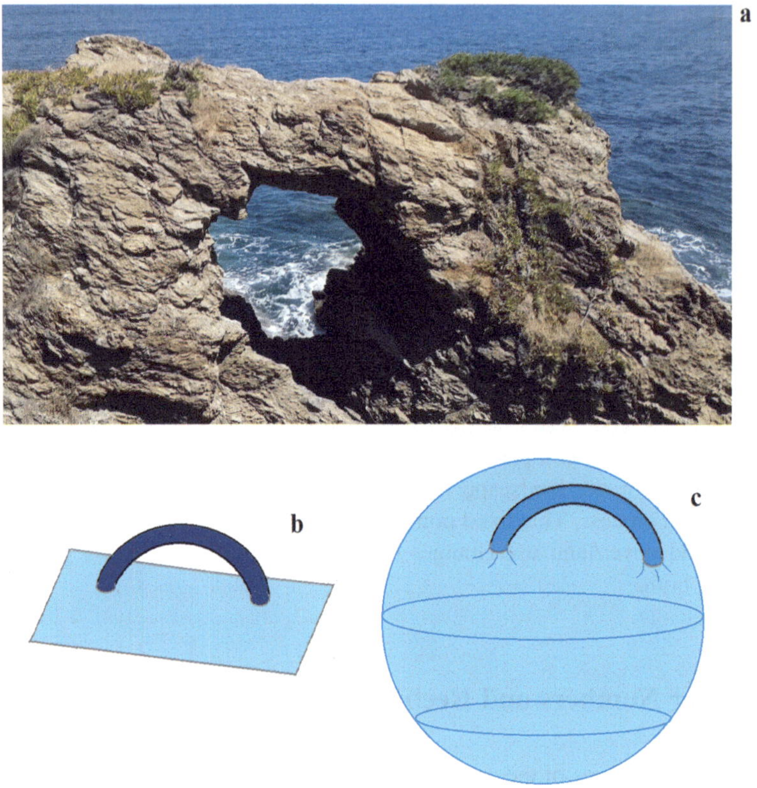

Fig. 6.5 A landform with a hole (**a**) in Galatas area in Greece, a handle fixed on the plane (**b**) and on the surface of a sphere (**c**)

strip, the cylinder and for the Klein bottle. Furthermore, it is also equal to the alternating sum of the Betti numbers of a surface:

$$\chi(S) = b_0 - b_1 + b_2 - b_3 \ldots$$

The genus of part of a surface of the earth need not always be zero; in fact, it may be simply connected or multiply connected (i.e. in the case of highway overpasses). In case a landform has only one hole (Fig. 6.5a), then one handle (Fig. 6.5b) corresponds to the cavity (Fig. 6.5c).

The calculation of Euler's number from the critical points (peaks, passes, pits) of the shape that is homeomorphic to the landform with hole is simple and straightforward and yields a confirmation of Euler's characteristic for tori: Peaks + Pits − Passes = 0 (Fig. 6.6).

The area of Alyki beach (Galatas area, central Greece) consists in a pair of beaches separated by a narrow strip of land that links two low hills. This area has

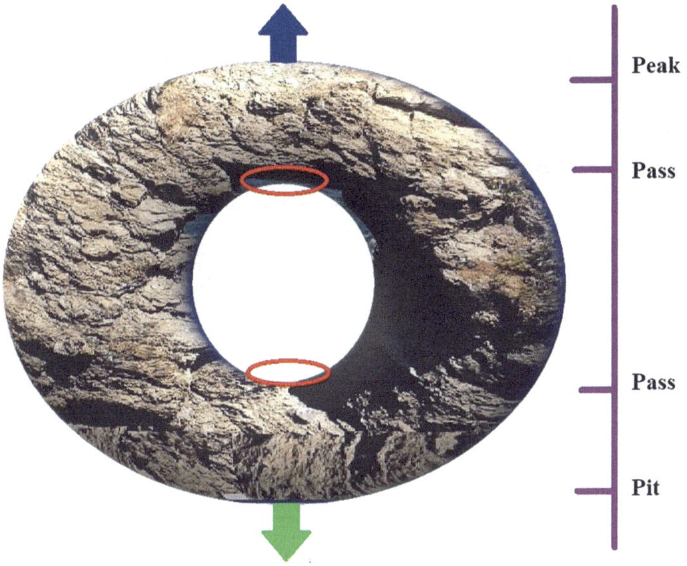

Fig. 6.6 The genus of the landform with a hole is zero and so is its Euler characteristic

two peaks, one virtual pit and a pass. Being homeomorphic to the sphere, its Euler characteristic is (Fig. 6.7):

$$\chi(S) = \text{Peaks} + \text{Pits} - \text{Passes} = 2 + 1 - 1 = 2$$

which is the Euler number of the sphere.

Euler numbers can be used on Reeb graphs which can be used to derive numerical descriptors of geometric shapes and forms (Tai et al., 1998; Biasotti et al., 2000; Cole-McLaughlin et al., 2004; Agarwal et al., 2006; Biasotti et al., 2008; Berretti et al., 2009; Morozov et al., 2013; Barra & Biasotti, 2013; Dey et al., 2013; De Silva et al., 2016). Having remained hitherto rather disregarded by geographers with few exceptions only (Pfaltz, 1976; Takahashi et al., 1995; Biasotti et al., 2004; Lukasczyk et al., 2015), Reeb graphs can be used to model and visualize topologically not only the geomorphic variations of the anaglyph but also the history of qualitative landscape changes. In the second case, this becomes possible by replacing "height" by "time" (see also Edelsbrunner et al., 2008).

A simplified map of an area of Kineta (Attica, 56 km west of Athens) appears suitable to illustrate the applicability of Reeb graphs: afforestations during the years 2012–2015 resulted in patches of new forests, although parts of the previously forested lands converted into shrublands during the same period. In the next 3 years (2015–2018), deforestation expanded further with one more patch of forest converted to shrubland (Fig. 6.8a).

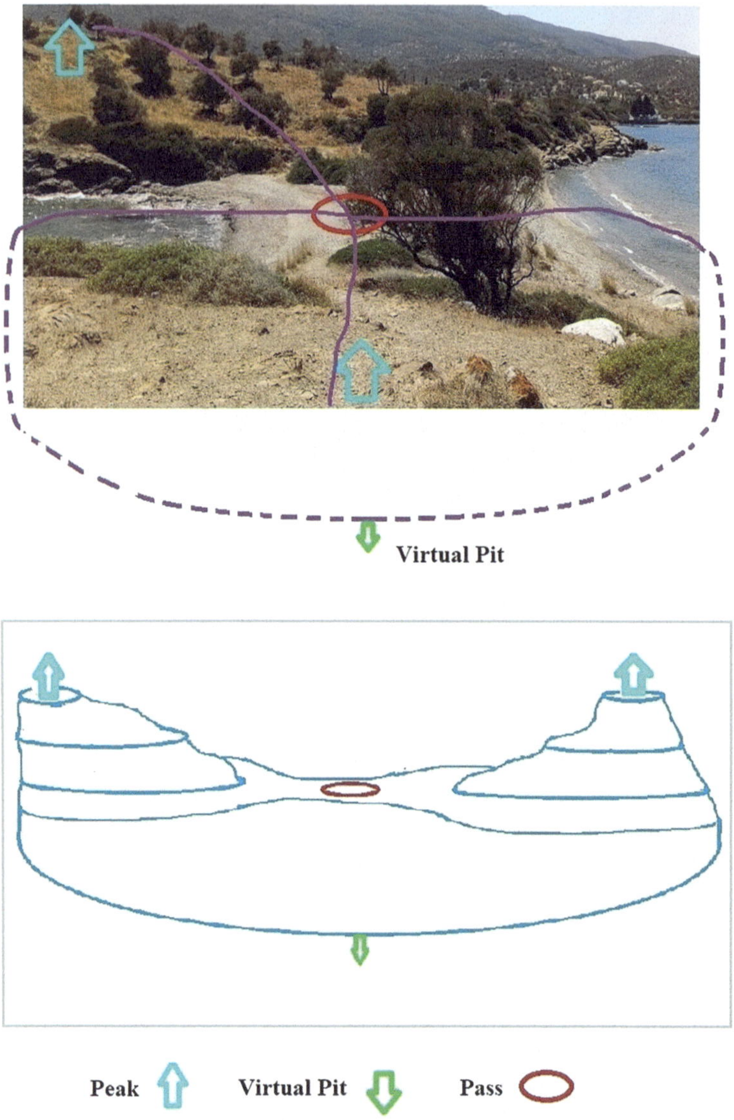

Fig. 6.7 Calculating the Euler number of the Alyki beach

These landscape changes can be described by homeomorphic shapes in which the "other" landscape type surrounds the forested areas, as the "landscape matrix" in landscape ecological terms (Fig. 6.8b).

Next, changes in tree cover (deforestation and afforestation) can be modelled by a Reeb graph in which the height function is time instead of topographic altitude (Fig. 6.9). From this graph, the Euler characteristic of the changes in forest areas is calculated as:

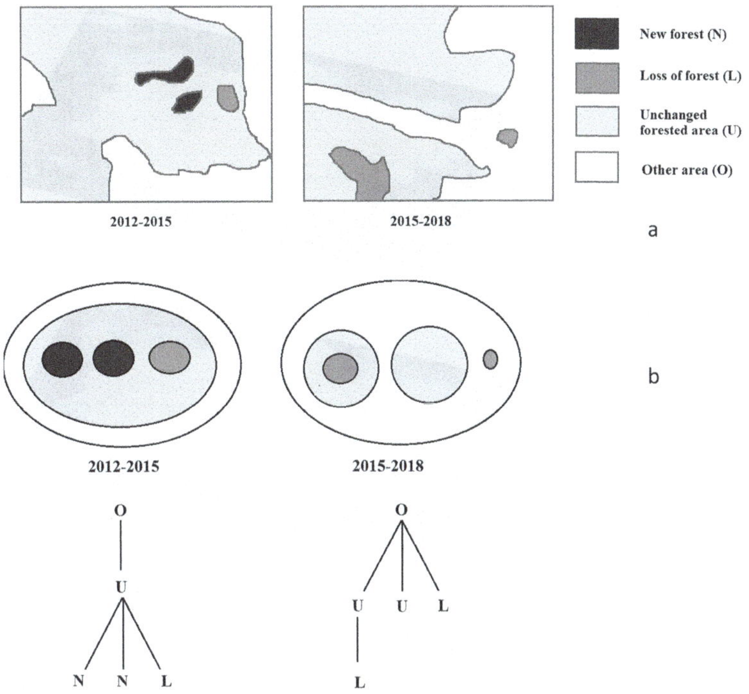

Fig. 6.8 The most important forested area changes in Kineta area from 2012 until 2018 (**a**) and their topological visualization using homeomorphic shapes and diagrams (**b**)

Fig. 6.9 The Reeb graph and the Euler characteristic of changes in the forested areas in Kineta, 2012 to 2022

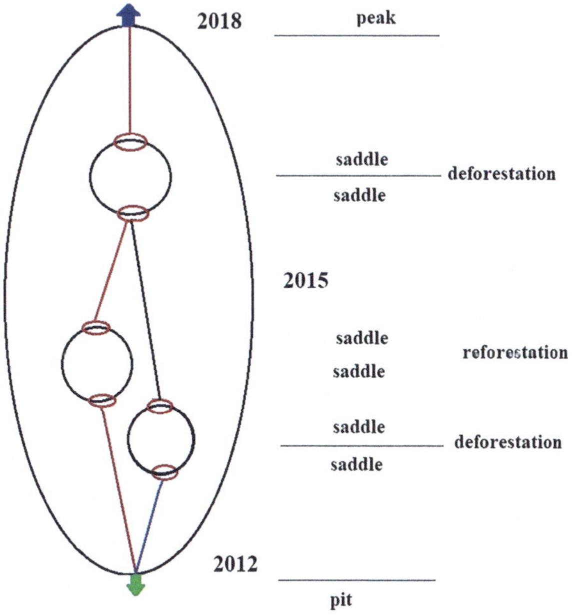

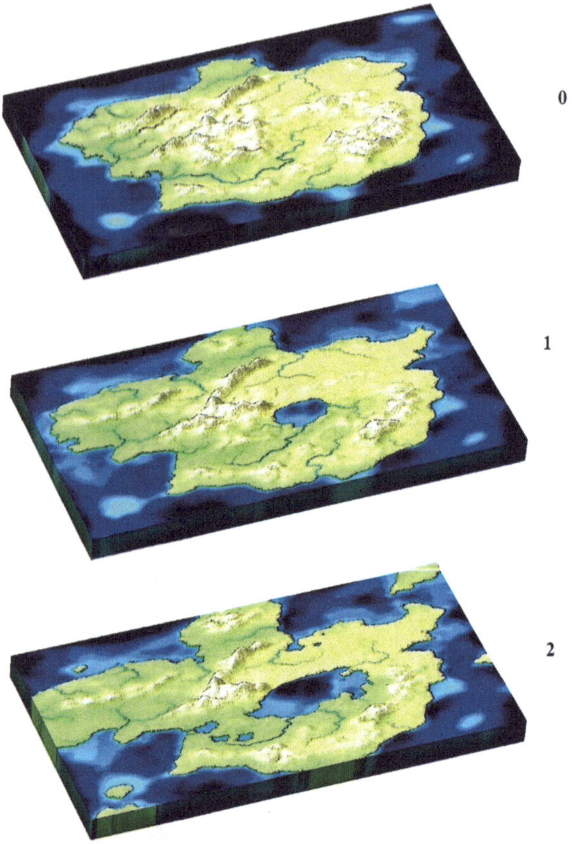

Fig. 6.10 Changes of a hypothetical island landscape from time 0 (top), through time 1 (middle) to time 2 (bottom)

$$\chi(S) = \text{Peaks} + \text{Pits} - \text{Passes} = 1 + 1 - 6 = -4$$

Long-term changes in geological time can be visualized in the same way. Consider for instance, the geomorphological changes of an artificial island landscape (Fig. 6.10) that has undergone geological changes (geological uplift, changes in the sea level etc) in the course of time and that these changes are depicted at time instants 0 (first), 1 (middle) and 2 (final).

These changes can be visualized by Reeb graphs from which it is also possible to calculate $\chi(S)$. If the calculation is carried out by considering the landscape change along a single timeline, then (Fig. 6.11a):

$$\chi(S)_{0\rightarrow 2} = \text{Peaks} + \text{Pits} - \text{Passes} = 1 + 1 - 8 = -6$$

But if $\chi(S)$ is calculated separately for each time interval, then two different calculations yield (Fig. 6.11b):

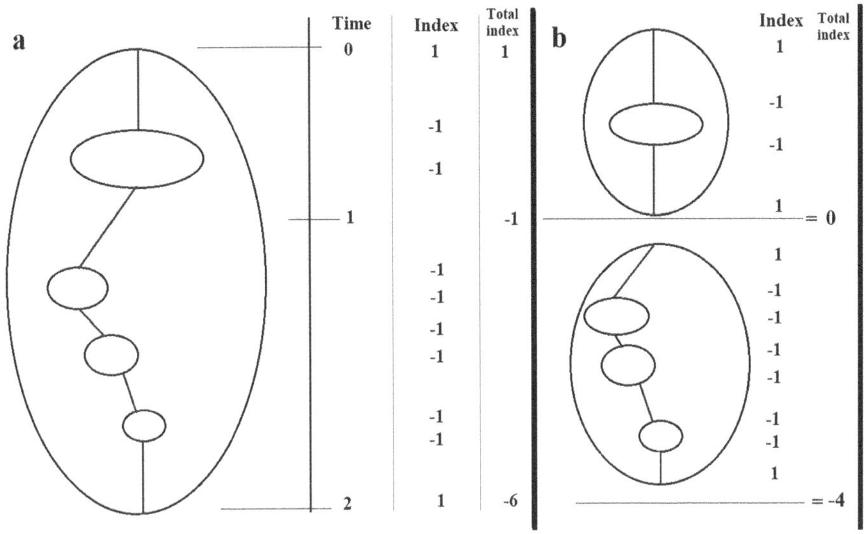

Fig. 6.11 The Reeb graph and calculations of critical points' indexes for the changes in the geomorphology of the island landscape in two different calculations: as a single timeline (**a**) and in two different timelines (**b**)

$$\chi(S)_{0\to 1} = 1 + 1 - 2 = 0$$
$$\chi(S)_{1\to 2} = 1 + 1 - 6 = -4$$

Hence, each history of landscape change is characterized by its own Reeb graph and concomitantly, its own Euler number.

References

Adelson-Velskii, G. M., & Kronrod, A. S. (1945). About level sets of continuous functions with partial derivatives. *Doklady Akademii Nauk SSSR, 49*(4), 239–241.

Agarwal, P. K., Edelsbrunner, H., Harer, J., & Wang, Y. (2006). Extreme elevation on a 2-manifold. *Discrete & Computational Geometry, 36*(4), 553–572.

Barra, V., & Biasotti, S. (2013). 3D shape retrieval using kernels on extended Reeb graphs. *Pattern Recognition, 46*(11), 2985–2999.

Berretti, S., Del Bimbo, A., & Pala, P. (2009). 3d mesh decomposition using Reeb graphs. *Image and Vision Computing, 27*(10), 1540–1554.

Biasotti, S., Falcidieno, B., & Spagnuolo, M. (2000, December 13–15). Extended Reeb graphs for surface understanding and description. In Discrete geometry for computer imagery: 9th international conference, DGCI 2000 Uppsala, Sweden. Proceedings 9 (pp. 185–197). Springer.

Biasotti, S., Falcidieno, B., & Spagnuolo, M. (2004). Surface shape understanding based on extended Reeb graphs. *Topological Data Structures for Surfaces: An Introduction for Geographical Information Science*, 87–103.

Biasotti, S., Giorgi, D., Spagnuolo, M., & Falcidieno, B. (2008). Reeb graphs for shape analysis and applications. *Theoretical Computer Science, 392*(1–3), 5–22.

Cole-McLaughlin, K., Edelsbrunner, H., Harer, J., Natarajan, V., & Pascucci, V. (2004). Loops in Reeb graphs of 2-manifolds. *Discrete & Computational Geometry, 32*(2), 231–244.

De Silva, V., Munch, E., & Patel, A. (2016). Categorified Reeb graphs. *Discrete & Computational Geometry, 55*(4), 854–906.

Dey, T. K., Fan, F., & Wang, Y. (2013). An efficient computation of handle and tunnel loops via Reeb graphs. *ACM Transactions on Graphics (TOG), 32*(4), 1–10.

Edelsbrunner, H., Harer, J., Mascarenhas, A., Pascucci, V., & Snoeyink, J. (2008). Time-varying Reeb graphs for continuous space–time data. *Computational Geometry, 41*(3), 149–166.

Lukasczyk, J., Maciejewski, R., Garth, C., & Hagen, H. (2015). Understanding hotspots: A topological visual analytics approach. In *Proceedings of the 23rd SIGSPATIAL international conference on advances in geographic information systems* (pp. 1–10).

Morozov, D., Beketayev, K., & Weber, G. (2013). Interleaving distance between merge trees. *Discrete and Computational Geometry, 49*(22–45), 52.

Pfaltz, J. L. (1976). Surface networks. *Geographical Analysis, 8*(1), 77–93.

Reeb, G. (1946). Sur les points singuliers d'une forme de Pfaff complètement intégrable ou d'une fonction numérique. *Comptes Rendues de l' Academie de Sciences de Paris, 222*, 847–849.

Tai, C. L., Shinagawa, Y., & Kunii, T. L. (1998). A Reeb graph-based representation for non-sequential construction of topologically complex shapes. *Computers & Graphics, 22*(2–3), 255–268.

Takahashi, S., Ikeda, T., Shinagawa, Y., Kunii, T. L., & Ueda, M. (1995). Algorithms for extracting correct critical points and constructing topological graphs from discrete geographical elevation data. In *Computer graphics forum* (Vol. 14, no. 3, pp. 181–192). Blackwell.

Chapter 7
Geo-topological Analysis of Land Use Dynamics

Topology is precisely the mathematical discipline dealing with the passage from the local to the global
(René Thom, 1923–2002, "Structural Stability and Morphogenesis")

7.1 Topology of Linear Dynamical Systems

Considering the problem of ecosystem stability in a simple form, May (1972, 2019) proposed a method of investigation of qualitative stability in model ecosystems with interactions of the type:

$$\frac{d\mathbf{X}(t)}{dt} = \mathbf{A}\mathbf{X}(t)$$

where **A** is a $n\text{x}n$ interaction matrix and $\mathbf{X}(t)$ is a $n\text{x}\mathbf{I}$ column matrix.

The necessary condition for this system to be stable is that the eigenvalues of the matrix **A** are all negative. However, the central theory for determining the long-term and qualitative behavior of a dynamical system modeled by a set of ordinary differential equations is the Lyapunov theory of stability (although not the only theory of stability), following which, a fixed point $\overline{x}$ is stable, if for any $\varepsilon > 0$ there exists a number $\delta(\varepsilon) > 0$ such that if $\|x_0 - \overline{x}\| < \delta$, then $\|x(t) - \overline{x}\| < \varepsilon$ for all $t > 0$. One of its main advantages lies in the fact that it is possible to derive general assessments of stability of an equilibrium point of a system without prior knowledge of the system's analytic solution (Papadimitriou, 2010).

In two-species linear systems (X and Y only), there are only six types of qualitative behaviors of the system, called "phase portraits" (Fig. 7.1):

(a) Unstable nodes (both eigenvalues are positive real)
(b) Stable nodes (both eigenvalues are negative real)
(c) Outward spirals (a pair of complex conjugate eigenvalues with positive real parts)
(d) Inward spirals (a pair of complex conjugate eigenvalues with negative real parts)

© The Author(s), under exclusive license to Springer Nature Switzerland AG 2023

F. Papadimitriou, *Geo-Topology*, GeoJournal Library 133,
https://doi.org/10.1007/978-3-031-48185-7_7

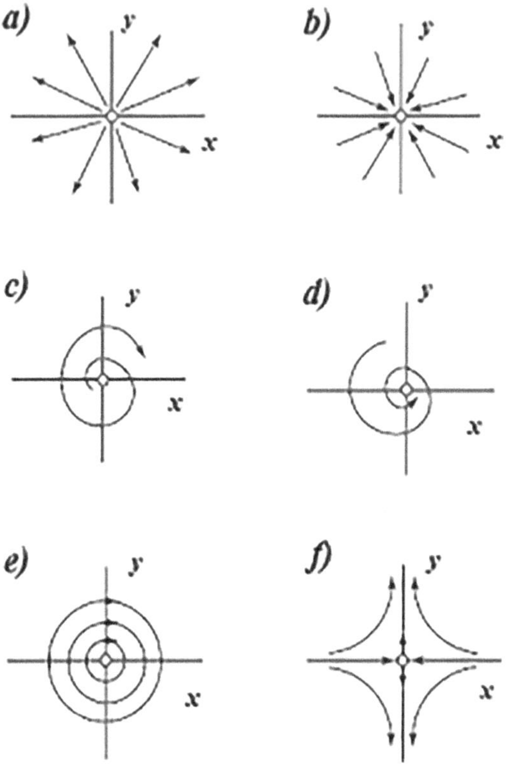

Fig. 7.1 Phase portraits of a 2D linear dynamical system: unstable node (**a**), stable node (**b**), outward spiral (**c**), inward spiral (**d**), centre (**e**), saddle point (**f**)

(e) Centres (both eigenvalues are purely imaginary numbers)
(f) Saddles (both eigenvalues are real, but one is positive and the other is negative).

Geographical dynamical systems of this type may contain two or three variables. Consider a linear model of three interacting land uses (*X*, *Y*, *Z*):

$$\frac{dX}{dt} = -aX + bY + cZ$$
$$\frac{dY}{dt} = aX - bY + hZ$$
$$\frac{dZ}{dt} = -(c+h)Z$$

The Jacobian matrix of this dynamical system is:

$$\begin{bmatrix} \frac{\partial X}{\partial X} & \frac{\partial X}{\partial Y} & \frac{\partial X}{\partial Z} \\ \frac{\partial Y}{\partial X} & \frac{\partial Y}{\partial Y} & \frac{\partial Y}{\partial Z} \\ \frac{\partial Z}{\partial X} & \frac{\partial Z}{\partial Y} & \frac{\partial Z}{\partial Z} \end{bmatrix} = \begin{bmatrix} -a & b & c \\ a & -b & h \\ 0 & 0 & -(c+h) \end{bmatrix}$$

The eigenvalues λ_i are calculated by solving for the determinant:

$$\det\left[\begin{pmatrix} -a & b & c \\ a & -b & h \\ 0 & 0 & -(c+h) \end{pmatrix} - \lambda\mathbf{I}\right] = \mathbf{0}$$

with **I** the unitary matrix.

The stability of the general linear dynamical system of three variables (X,Y,Z) depends on the eigenvalues λ_i of the system's Jacobian, and hence on the characteristic polynomial of third degree that is derived from the Jacobian's determinant. For a 3D linear system such as this, the general form of its cubic equation is:

$$\lambda^3 + p\lambda^2 + q\lambda + r$$

where

$$p = -(a+b+c+h)$$

and

$$q = -(ac+bc+bh+ah)$$

Defining the parameter δ of the cubic equation as:

$$\delta = pq - r$$

and the discriminant Δ of the characteristic third-degree polynomial:

$$\Delta = -p^2q^w + 4p^3r + 4q^3 + 27r^2 - 18pqr$$

The stability types of the system are thus described according to the values of Δ, δ, r and q (Table 7.1) and correspond to some typical phase portraits (Fig. 7.2).

7.2 Topological Aspects of Structural Dynamics

Landscape transformations result in re-arrangements of the topology of landscape patches, so while some boundaries among different land use patches are lost, new ones emerge. Although the sizes of land patches that changed land use vary, it is reasonable to ask what is the relative *frequency* of their transformations, or, equivalently: "How many times does a given land use transformation appear in separate patches?". So the number of patches of each land use transformation has to be counted from land use change maps (how many times a land use transformation occurs as a separate patch).

Table 7.1 The stability types of 3D linear dynamical systems

Equilibrium point	Parameters	Discri-minant	Eigenvalues λ_i
Stable node	$\delta > 0, r > 0, q > 0$	$\Delta < 0$	$\text{Im}\lambda_j = 0, \lambda_j < 0$
Stable focus	$\delta > 0, r > 0, q > 0$	$\Delta > 0$	$\text{Re}\lambda_{1,2} < 0, \lambda_3 < 0$
Unstable node	$\delta < 0, r < 0, q > 0$	$\Delta > 0$	$\text{Im}\lambda_j = 0, \lambda_j > 0$
Unstable focus	$\delta < 0, r < 0, q > 0$	$\Delta < 0$	$\text{Re}\lambda_{1,2} > 0, \lambda_3 > 0$
Saddle node	$\delta > 0, r < 0, q <= 0$ $\delta > 0, r < 0, q > 0$ $\delta < 0, r > 0, q <= 0$ $\delta < 0, r > 0, q > 0$	$\Delta < 0$ $\Delta < 0$ $\Delta < 0$ $\Delta < 0$	$\text{Im}\lambda_j = 0, \lambda_{1,2} < 0, \lambda_3 > 0$ $\text{Im}\lambda_j = 0, \lambda_{1,2} < 0, \lambda_3 > 0$ $\text{Im}\lambda_j = 0, \lambda_{1,2} > 0, \lambda_3 < 0$ $\text{Im}\lambda_j = 0, \lambda_{1,2} > 0, \lambda_3 < 0$
Saddle focus	$\delta > 0, r < 0, q <= 0$ $\delta > 0, r < 0, q > 0$ $\delta < 0, r > 0, q <= 0$ $\delta < 0, r > 0, q > 0$	$\Delta > 0$ $\Delta > 0$ $\Delta > 0$ $\Delta > 0$	$\text{Re}\lambda_{1,2} < 0, \lambda_3 > 0$ $\text{Re}\lambda_{1,2} < 0, \lambda_3 > 0$ $\text{Re}\lambda_{1,2} > 0, \lambda_3 < 0$ $\text{Re}\lambda_{1,2} > 0, \lambda_3 < 0$

A topological approach is useful to examine the dynamics of these changes and the region of East Attica may be a suitable case (an 850 km^2 large area east of Athens, Greece), of which the land use transformations can be evaluated (Papadimitriou, 1997) from the overlay of the map of land use types of 1988 over the map of 1967 (Fig. 7.3).

The map overlay yields 155 patches of land use transformations that can be classified in a transitions matrix:

$$\begin{pmatrix} & U & F & A & S & B & O \\ U & 0 & 0 & 0 & 0 & 0 & 0 \\ F & 10 & 0 & 10 & 24 & 8 & 7 \\ A & 17 & 7 & 0 & 13 & 9 & 3 \\ S & 3 & 14 & 7 & 0 & 9 & 5 \\ B & 1 & 2 & 2 & 4 & 0 & 0 \\ O & 0 & 0 & 0 & 0 & 0 & 0 \end{pmatrix}$$

It is thus possible to calculate the change in patchiness for each land use type in the course of time for the time step that corresponds to the cartographic information that is available (21 years in this case). To simplify calculations however, the total numbers of patches for U_{BO}, F, A, S are re-calculated as U_{BO}-F-A-S (where U_{BO} is the sum of values of the entries of U, B and O):

$$\begin{pmatrix} & U_{BO} & F & A & S \\ U_{BO} & 0 & 1 & 2 & 4 \\ F & 25 & 0 & 11 & 24 \\ A & 30 & 7 & 0 & 13 \\ S & 7 & 14 & 7 & 0 \end{pmatrix}$$

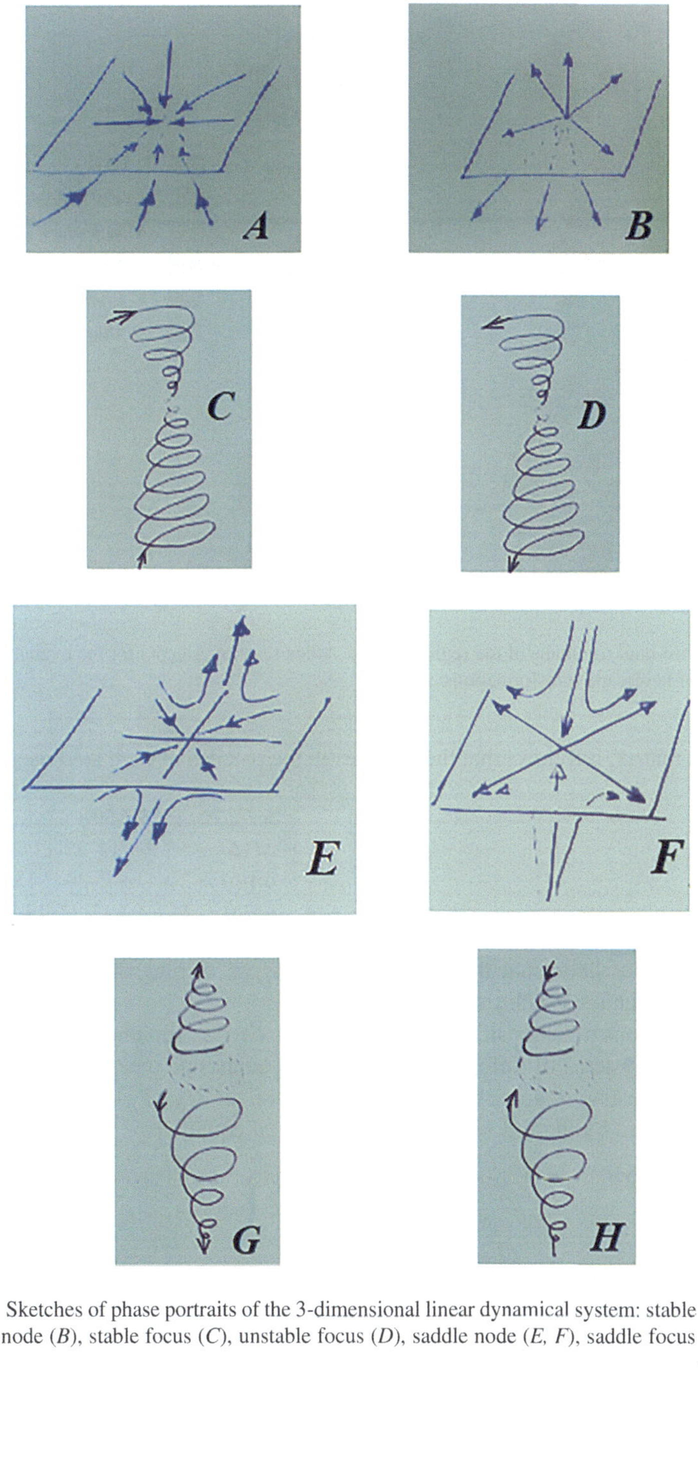

Fig. 7.2 Sketches of phase portraits of the 3-dimensional linear dynamical system: stable node (*A*), unstable node (*B*), stable focus (*C*), unstable focus (*D*), saddle node (*E, F*), saddle focus (*G, H*)

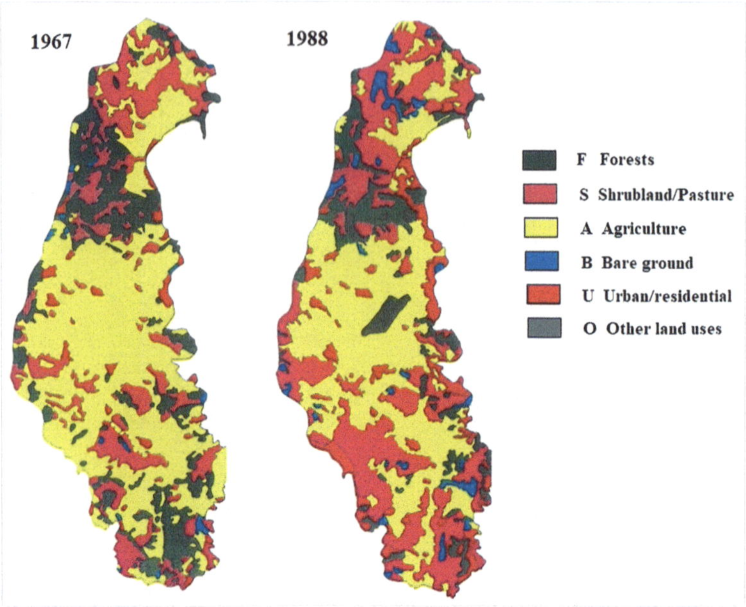

Fig. 7.3 Two land use maps of the region of East Attica (east of Athens) for the evaluation of the dynamics of landscape transformations

From this matrix, it follows that the changes in the patches of each land use type are:

U_{BO}	$\Delta U_{BO}/\Delta t = 25 + 30 + 17–1–2-4 = 65$
F	$\Delta F/\Delta t = 1 + 7 + 14–25–11–24 = -38$
A	$\Delta A/\Delta t = 2 + 11 + 7–30–7-13 = -30$
S	$\Delta S/\Delta t = 24 + 4 + 13–17–14-7 = 3$

These results show that the U_{BO} landscape types display the highest rates of increase in patchiness with time.

The transitions of land patches from one land use type to another can be modelled by 2D or 3D systems of differential equations of landscape types. Thus, beginning with the simplest approach which is a linear system, the U_{BO}-F-A system of patches can be modelled by the following set of differential equations:

$$\frac{dU_{BO}}{dt} = -3U_{BO} + 25F + 30A$$
$$\frac{dF}{dt} = U_{BO} - 36F + 7A$$
$$\frac{dA}{dt} = 2U_{BO} + 11F - 37A$$

The qualitative dynamics of this system in the phase space is derived from the calculation of eigenvalues from its Jacobian:

$$\mathbf{J}(U_{BO}, F, A) = \begin{pmatrix} \frac{\partial U_{BO}}{\partial U_{BO}} & \frac{\partial U_{BO}}{\partial F} & \frac{\partial U_{BO}}{\partial A} \\ \frac{\partial F}{\partial U_{BO}} & \frac{\partial F}{\partial F} & \frac{\partial F}{\partial A} \\ \frac{\partial A}{\partial U_{BO}} & \frac{\partial A}{\partial F} & \frac{\partial A}{\partial A} \end{pmatrix}$$

and thus the eigenvalues are calculated from the equation

$$\det(\mathbf{J}(U_{BO}, F, A) - \lambda \mathbf{I}) = \mathbf{0}$$

which yields the eigenvalues $\lambda_1 = -45.41$, $\lambda_2 = -30.58$, $\lambda_3 = 0$ and thus the system is "degenerate" and its topology is uninteresting. By the same procedure, it can be proven that all the remaining 3 × 3 landscape patch subsystems have the same topological characteristics and configuration in the phase space.

But if non-linear interactions were introduced to model the systems of patch dynamics, then more interesting topologies would emerge from the more complex behaviors that would be revealed. An appropriate nonlinear such model of patch dynamics would be a Lotka-Volterra, which can be created as follows i.e. for the *F*-*S* subsystem:

Considering that the there were 58 patches of *F* and 80 patches of *S* initially, the rate of growth of forests α=14/58=0.241 and the rate of depletion of patches of *S* is γ =24/80=0.3, while the rate of depletion of *S* due to the interaction with *F* is β=14/80=0.175 and the rate of depletion of forests due to the expansion of shrublands is δ =24/(58+80) =0.174. Thus, the non-linear model is

$$\frac{dF}{dt} = \alpha F - \beta FS = 0.241F - 0.175FS$$
$$\frac{dS}{dt} = \delta FS - \gamma S = 0.174FS - 0.3S$$

The steady states are at

$$(dF/dt, dS/dt) = (0,0) \text{ and} (dF/dt, dS/dt) = (\gamma/\delta, \alpha/\beta) = (1.764, 1.377).$$

At the steady state (0,0) the Jacobian is

$$J(0,0) = \begin{bmatrix} \alpha & 0 \\ 0 & -\gamma \end{bmatrix}$$

and thus the eigenvalues are $\lambda_1 = \alpha = 0.241$ and $\lambda_2 = -\gamma = -0.3$, which means that the extinction of both *F* and *S* is impossible.

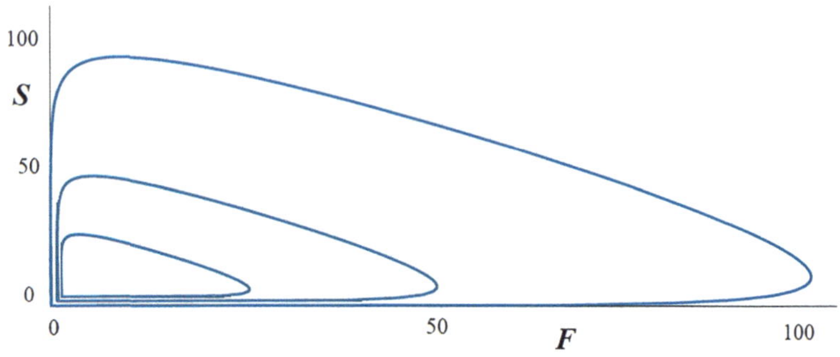

Fig. 7.4 Orbits of the system *F*-*S* around the fixed point with the oscillating behavior

At the steady state (1.764, 1.377) the Jacobian is

$$J(1.764, 1.377) = \begin{bmatrix} 0 & -\frac{\beta\gamma}{\delta} \\ \frac{\alpha\delta}{\beta} & 0 \end{bmatrix}$$

The eigenvalues are calculated as

$$\lambda_1 = i\sqrt{\alpha\gamma} = 0.269i$$
$$\lambda_2 = -i\sqrt{\alpha\gamma} = -0.269i$$

and define a fixed point around which the orbits of the *F*-*S* system oscillate (Fig. 7.4) with a period equal to

$$\mathrm{T} = \frac{2\pi}{\sqrt{\lambda_1\lambda_2}}$$

Further, some different insights into the patch dynamics can be gained by considering the divergence of a vector field V in the R^m (m-dimensional) Euclidean space with Cartesian coordinates X_{i}:

$$div\mathbf{F} = \nabla\mathbf{F}$$

for a linear system (equivalent to a vector field) with the general form.

$$V(X) = \mathbf{A}^* X,$$

where the divergence is equal to the trace of the matrix **A**.

The number of patches defines whether the topology of the landscape is "coarse" or "fine" and the change in this number reflects the long-term changes in the topology of a landscape. In turn, this "dynamics" can be assessed in terms of energy dissipation. A measure of energy dissipation in a dynamical system can be its divergence. If $divV = 0$ (the trace of the Jacobian of the matrix **A** is zero), the phase flow preserves volumes of any region of the phase space and hence there is neither waste nor gain of energy by the system. In dissipative systems, energy changes over time, whilst in non-dissipative (Hamiltonian) systems the energy is constant. Mechanical systems that are described by differential equations with negative divergence (negative trace of the Jacobian) are characterized by friction and dissipation of energy. The relative growth or shrinkage of the area of the phase space (with respect to its initial conditions) is determined by the partial derivatives of the time evolution (differential) equations of the dynamical system. If the trace of the Jacobian is negative, then the initial phase space area shrinks, the system is dissipative, the system's trajectories concentrate to an attractor and that attractor is of a lower dimension in comparison to that of the original state space. Otherwise put, negative divergence means contraction of volume in the phase space of a system and, consequently, energy dissipation.

Of all the 2D land use sub-systems of East Attica, the *F-S* system has the highest energy dissipation. To verify this, it suffices to show that the trace of the matrix of *F-S* has the lowest value in comparison to all the traces of the matrices of the two dimensional systems (*F-A*, *A- S*, U_{BO}-*A*, U_{BO}-*S*, U_{BO}-*F* and *F-S*). Indeed, the 2D patchiness systems of the landscape's structural dynamics are:

$\frac{dF}{dt} = -11F + 7A$ $\frac{dA}{dt} = 11F - 7A$	$\frac{dF}{dt} = -24F + 14S$ $\frac{dS}{dt} = 24F - 14S$
$\frac{dA}{dt} = -13A + 7S$ $\frac{dS}{dt} = 13A - 7S$	$\frac{dU_{BO}}{dt} = -2U_{BO} + 30A$ $\frac{dA}{dt} = 2U - 30A$
$\frac{dU_{BO}}{dt} = -4U_{BO} + 17S$ $\frac{dS}{dt} = 4U_{BO} - 17S$	$\frac{dU_{BO}}{dt} = -U_{BO} + 25F$ $\frac{dF}{dt} = U_{BO} - 25F$

so the traces of their Jacobian matrices are:

System	Trace
F-S	−38
U_{BO}-*A*	−32
F-A	−18
A-S	−20
U_{BO}-*F*	−26
U_{BO}-*S*	−21

Similarly, of all the 3D systems, the U_{BO}-*F*-*S* has the highest energy dissipation:

System	Trace
U_{BO}-*F*-*S*	−85
F-*A*-*S*	−76
U_{BO}-*A*-*S*	−73
U_{BO}-*F*-*A*	−76

and, as a matter of fact, the U_{BO} -*F*-*S* patch system converges to an attractor faster than any other 3D landscape system of East Attica.

It can be observed that the *F*-*A*-*S* system ranks second in its relative dissipation rate in 3D systems and first in 2D systems, so the dynamics of this subsystem concentrates some of the most ecologically important characteristics of the region. From the results obtained for both the 2D and 3D patch systems, it can be observed that in three out of four cases, the maximum dissipation is associated with land use types involving the *F*-*S* combination. This means that the forest-shrubland pair constitutes the key-system with respect to *energy dissipation* in East Attica. Otherwise put, the landscape changes involving the pair forest-shrubland lead to an attractor quicker than any other combination of landscape types of this region. Yet, considering the wide expansion of forest-shrubland dynamics in space and time leading to land degradation in Mediterranean-type ecosystems (Papadimitriou & Mairota, 1996), this kind of dynamics is probably not restricted to the region of East Attica only and might well characterize other Mediterranean-type ecosystems as well.

7.3 Applying the Poincaré-Bendixon Theorem

Another fixed point theorem that highlights the usefulness of topology in the study of long term land use changes is the Poincaré-Bendixon theorem, which assures that any bounded solution of a differential equation in the plane either converges to a fixed point or to a closed orbit (a limit cycle), thus ruling out the possibility of chaotic behaviors in the plane. The theorem does not hold for dimensions equal or higher than three (in these spaces the system's trajectories may wander forever without converging to some attractor, fixed point or orbit), but it can be applied to modelling land use dynamics by using a global descriptor of land use change. One such is "landscape entropy" or "landscape diversity" (Papadimitriou & Mairota, 1998; Papadimitriou, 2022a, b, c, d) as it summarizes both the number and the relative participation of land use types in a landscape at any moment and is measured from the Shannon formula:

$$H = -\sum_{i=1}^{i=n} P_k \ln P_k$$

where P_k stands for the percentage of each land use ($i = 1,2,3,\ldots k$) of the landscape.

Consider the data on agricultural land use of Slovenia, based on the country's national statistics from 1929 until 1984 (see Vrišer, 1990, 1993; Gabrovec & Kumer, 2019). These data can be used to calculate the entropy of the agricultural land use of Slovenia for that period by using the Shannon formula for the country's official records of the $k = 7$ agricultural land use types that have been registered in the country's statistics: cereals, vegetables, fodder cultivations, grassland, orchards, vineyards and hop-fields (all types of agricultural landscapes of the country are taken into account). Changes of entropy with time can be indicative of land degradation and/or landscape homogeneization in Southern Europe (Papadimitriou & Mairota, 1996, 1998a, b).

After the calculations, the phase portrait of the agricultural landscapes of the country can be represented in the phase space that is defined by H and dH/dt. The changes in the entropy of land use types are continuous because landscapes change continuously with time (the fact that entropy was sampled at certain discrete time instants does not mean that the process of land use entropy change is discrete itself). Consequently, a question arises as to the interpretation of the graphic representation of the dynamics in the phase space, which, put differently, is tantamount to asking "what insights might H and dH/dt give us about this landscape's long-term changes?"

As can be easily verified, the graphic representations of H against time or dH/dt against time offer no deeper insight into the land use dynamics. But plotting H against dH/dt in the phase space can be quite revealing, since a characteristic pattern emerges if H is plotted against dH/dt and the difference equations are approximated by a continuous curve (Fig. 7.5).

Topologically, this curve is a "Poincaré mapping" or "monodromy" or "holonomy" of the dynamical system of land use types, describing the qualitative behaviour of land uses in the phase space. The equation that has exactly this Poincaré mapping is the equation of an harmonic oscillator with damping:

$$\frac{d^2H}{dt^2} + \mu\frac{dH}{dt} + H = 0, \text{ with } \mu > 0.$$

The critical point of this equation is at $H = dH/dt = 0$, corresponding to an equilibrium solution. In fact, this phase portrait matches the *Liénard* equation:

$$\frac{d^2H}{dt^2} + \mu\frac{dH}{dt} + H - 1.136 = 0$$

which is a special case of the generalised *Liénard* equation:

$$\frac{d^2H}{dt^2} + \mu\frac{dH}{dt} + g(H) = 0.$$

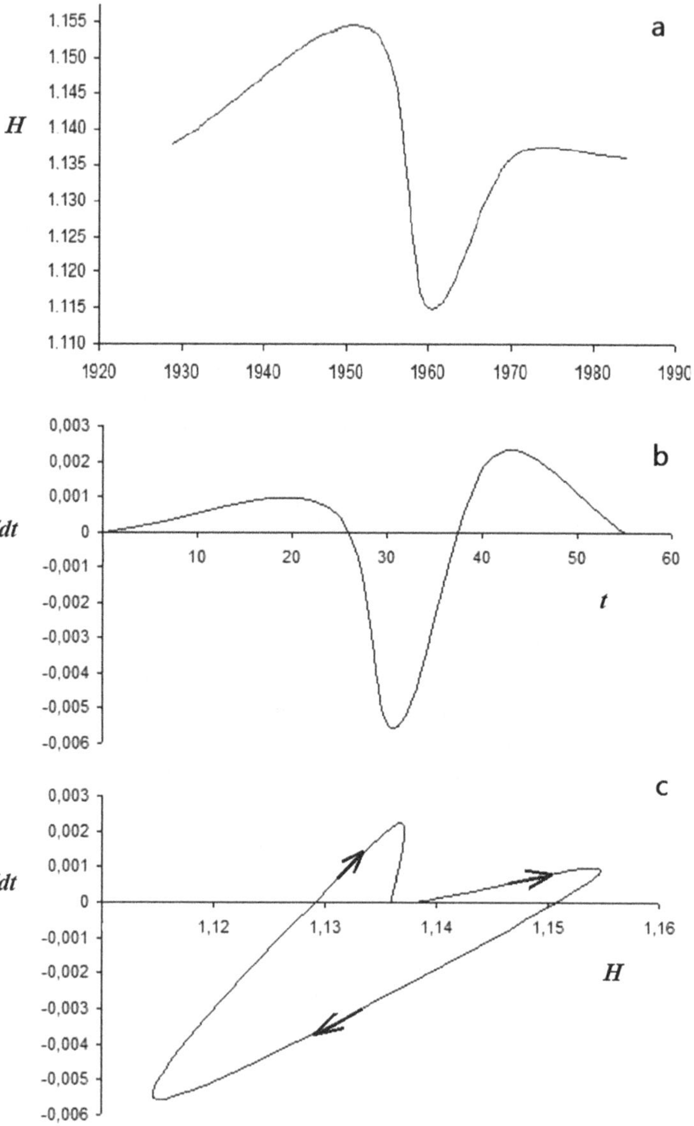

Fig. 7.5 The entropy changes of the agricultural land uses of Slovenia from 1929 until 1984 (**a**), the mean annual entropy change (**b**), the change with respect to time (from 1929 onwards) and the curve (**c**) showing the change of H against dH/dt

The *Liénard* equation has its only critical point at (0,0). Then, by virtue of the *Poincaré-Bendixon theorem* it is guaranteed that there is at least one periodic solution of this equation (Sabatini, 2004; Atslega & Sadyrbaev, 2013). Considering that a T-periodic solution $\varphi(t)$ of the *Liénard* equation is stable if

$$\int_0^{\tau} \varphi(\tau)dt > 0,$$

and T-periodic solutions of the *Liénard* equation can be handled in a way similar to the equation

$$\frac{dx}{dt} = \mathbf{A}(t)x,$$

where $\mathbf{A}(t)$ is a nxn matrix, it is concluded that the long-term change in the entropy of the agricultural land use of Slovenia has a stable periodic solution if $\lambda_1 = 0$ and $\lambda_2 \leq 0$, where λ_i are exponents of the previous equation and

$$\lambda_2 = -\frac{1}{\mathrm{T}} \int_0^{\mathrm{T}} f(\varphi(\tau))dt \mathrm{mod}\left(\frac{2\pi i}{\mathrm{T}}\right)$$

Thus, the changes in the entropy of the relative participations of various agricultural land uses in Slovenia can be studied in a phase space that is defined by entropy (H) and entropy change (*dH*/*dt*), while the holonomy or "Poincaré mapping" of the landscape's behaviour can be approximated by a generalized *Liénard* equation, which, as the topology of its phase space reveals the sign of periodicity.

References

Atslega, S., & Sadyrbaev, F. (2013). On solutions of Liénard type equations. *Mathematical Modelling and Analysis, 18*(5), 708–716.

Gabrovec, M., & Kumer, P. (2019). Land-use changes in Slovenia from the Franciscean cadaster until today. *Acta Geographica Slovenica, 59*(1), 64–81.

May, R. M. (1972). Will a large complex system be stable? *Nature, 238*(5364), 413–414.

May, R. M. (2019). *Stability and complexity in model ecosystems*. Princeton University Press.

Papadimitriou, F. (1997). *Land use modelling, land degradation and land use planning in East Attica, Greece*. University of Oxford.

Papadimitriou, F. (2010). Mathematical modelling of spatial-ecological complex systems: An evaluation. *Geography, Environment, Sustainability, 1*(3), 67–80.

Papadimitriou, F. (2022a). *Spatial entropy and landscape analysis*. Springer.

Papadimitriou, F. (2022b). Spatial entropy geo-information and spatial surprise. In *Spatial entropy and landscape analysis* (pp. 1–15). Springer.

Papadimitriou, F. (2022c). Computing the spatial entropy of square binary maps. In *Spatial entropy and landscape analysis* (pp. 15–30). Springer.

Papadimitriou, F. (2022d). Spatial entropy, non-extensive thermodynamics and landscape change. In *Spatial entropy and landscape analysis* (pp. 103–121). Springer.

Papadimitriou, F., & Mairota, P. (1996). Spatial scale-dependent policy planning for land Management in Southern Europe. *Environmental Monitoring and Assessment, 39*(1–3), 47–57.

Papadimitriou, F., & Mairota, P. (1998a). Land use diversity changes. In P. Mairota, J. Thornes, & N. Geeson (Eds.), *Atlas of Mediterranean environments in Europe* (pp. 89–90). Wiley.
Papadimitriou, F., & Mairota, P. (1998b). Agriculture. In P. Mairota, J. Thornes, & N. Geeson (Eds.), *Atlas of Mediterranean environments in Europe* (pp. 86–89). Wiley.
Sabatini, M. (2004). On the period function of $x''+f(x)x'^2+g(x)=0$. *Journal of Differential Equations,* 196(1), 151–168.
Vrišer, I. (1990). *Agrarian systems in the Republic of Slovenia* (pp. 53–60). Geographica Iugoslavica.
Vrišer, I. (1993). Agrarian economy in Slovenia. *Geo Journal, 31*, 373–377.

Chapter 8
Geo-topological Visualization with Knots and Braids

While the scientist sees everything that happens in one point of space, the poet feels everything that happens in one point of time
(Vladimir Nabokov, 1899–1977, "Speak Memory")

8.1 Braid Models of Landscape Change

Braids are mathematical objects consisting of interlacing strands. A braid of n strands is a set of pairwise nonintersecting ascending strands joining the points of one plane ($A_1, \ldots, A_n$) to the points of another, parallel, plane ($B_1, \ldots, B_n$) in the space R^3. A braid can be described algebraically by series of symbols, the "braid words" (Artin, 1947; Markoff, 1945; Crowell & Fox, 1963; Gordon, 1978; Kauffman, 1987; Hemion, 1992).

Geographers are not unfamiliar with braiding; river braiding is a recurrent theme in geomorphological research (Foufoula-Georgiou & Sapozhnikov, 2001; Piégay et al., 2006; Williams et al., 2016), but the topological object "braid" was introduced only very recently (Papadimitriou, 2020) to geographical analysis.

Consider an example landscape with four land uses (F, A, S, P) and suppose that this landscape changed in the course of time with some of its areas transformed from one land use to another (Fig. 8.1).

Landscape transformations are registered in transition matrices in which the rows represent conversion "from" and columns "to". An entry 1 indicates that a land use type of the landscape converted to another land use type. If the entry is 0, then that land use type has remained unchanged during the time step. Thus, from time 0 to time 1 the changes are:

© The Author(s), under exclusive license to Springer Nature Switzerland AG 2023

F. Papadimitriou, *Geo-Topology*, GeoJournal Library 133,
https://doi.org/10.1007/978-3-031-48185-7_8

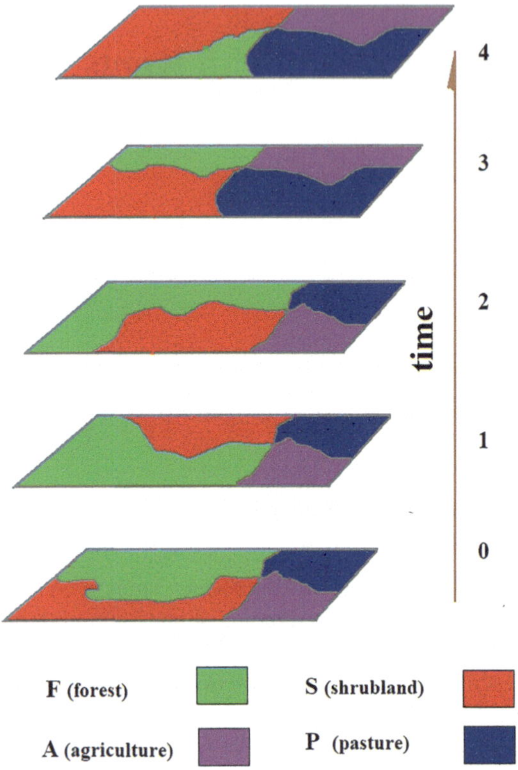

Fig. 8.1 An example landscape with four land uses and conversions from one land use to another in the course of four time steps

$$\begin{pmatrix} & F & P & S & A \\ F & 0 & 0 & 1 & 0 \\ P & 0 & 1 & 0 & 0 \\ S & 0 & 0 & 0 & 0 \\ A & 0 & 0 & 0 & 1 \end{pmatrix}$$

and from time 1 to time 2 the transitions are

$$\begin{pmatrix} 0 & 0 & 0 & 0 \\ 0 & 1 & 0 & 0 \\ 1 & 0 & 0 & 0 \\ 0 & 0 & 0 & 1 \end{pmatrix}.$$

Similarly, from time 2 to time 3:

$$\begin{pmatrix} 1 & 0 & 0 & 0 \\ 0 & 0 & 0 & 0 \\ 0 & 0 & 1 & 0 \\ 0 & 1 & 0 & 0 \end{pmatrix}$$

and, eventually, from time 3 to 4: $_\nu$

$$\begin{pmatrix} 1 & 0 & 0 & 0 \\ 0 & 0 & 0 & 0 \\ 0 & 0 & 1 & 0 \\ 0 & 1 & 0 & 0 \end{pmatrix}.$$

This algebraic description of land use changes makes topology-based visualizations possible that can be created by using braid theory (Papadimitriou, 2020): whenever a land use type is transformed into another, then a braid can be used to visualize the case that the "beneath" strand of that land use is transformed into the land use type represented by the "above" strand. For a braid with four strands (with each strand corresponding to a different land use type), there are some basic interlacing patterns, depending on whether one strand is woven above or under another (Fig. 8.2).

Hence, a braid model can be created that offers a visualization of the land conversions on the example landscape (Fig. 8.3). Consequently, the entire process of this landscape change corresponds to the *braid word*:

$$b_1 b_1{}^{-1} b_3 b_1{}^{-1}$$

But braids may not only model the long-term landscape changes from one land use type to another, for they might as well be used to model and visualize the transformations that take place *within* a patch (Fig. 8.4). Expectedly, the braid word that would describe that landscape transformation process is different in this case:

$$b_1{}^{-1} b_2{}^{-1} b_1 b_1{}^{-1} b_1{}^{-1} b_2 b_1{}^{-1} b_3$$

and corresponds to a different braid visualization (Fig. 8.5).

Braid models may also be indicative of the complexity of the object modeled (Kholodenko & Rolfsen, 1996; Diao & Ernst, 1998; Fiedler & Rocha, 1999; Hamidi-Tehrani, 2000; Nechaev et al., 1996; Barenghi et al., 2001; Bangert et al., 2002; Garber et al., 2002; Makowsky & Marino, 2003; Nechaev & Voituriez, 2005; Orlandini et al., 2005; Thiffeault, 2005; Dynnikov & West, 2007) and thus of the complexity of landscape changes also (Papadimitriou, 2020). This is based on the understanding that the minimum number of crossings of a braid is a measure of braid

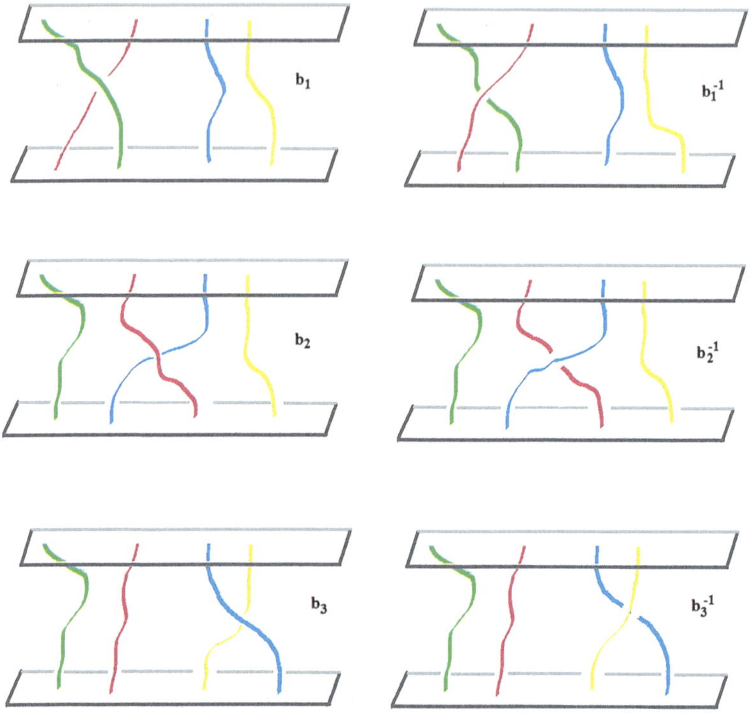

Fig. 8.2 The elementary braids consisting of four strands

complexity (Berger, 1994; Bangert et al., 2002; Simsek et al., 2003) which can also serve as a measure of landscape complexity (Papadimitriou, 2020).

8.2 Knot Models of Landscape Change

A knot is a closed curve embedded in the 3D space, a topological embedding of a circle in 3D Euclidean space (Kassel & Turaev, 2008). Knots can be added by cutting them and then splicing their ends together without making any new crossings. Much like braids have their characteristic words, knots have their characteristic polynomials (Kawauchi, 1996; Lickorish, 1997; Murasugi, 2007). Also like braids, knots can serve as models for the visualization of long-term land use changes.

Consider, for instance, a relatively simple knot, the "*cinquefoil*" (the knot 5_1). For a landscape with five land use types (F, A, S, P, B) and their transformations, a cinquefoil is a suitable topological model to visualize these transformations (Fig. 8.6). At each one of its crossings, the "under" strand corresponds to the land use type that is transformed to the land use type that corresponds to the "above"

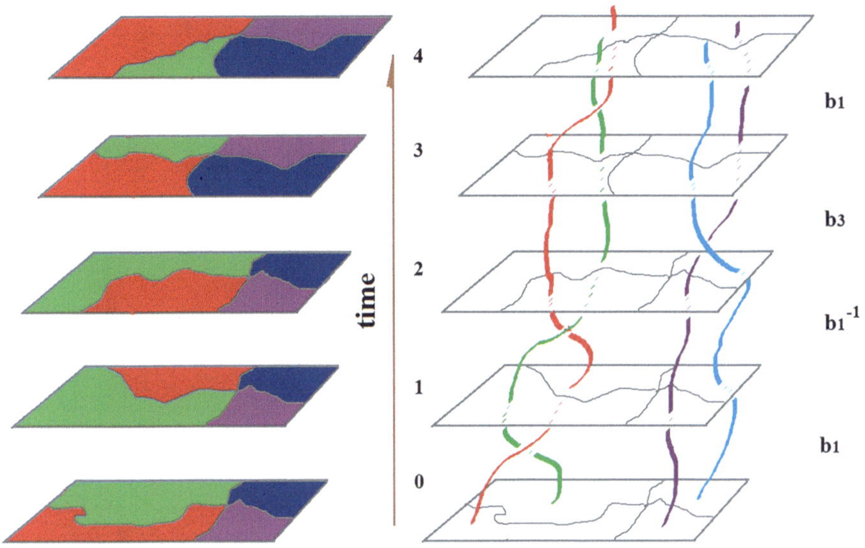

Fig. 8.3 A braid model of landscape changes, depicting transformations to and from a land use type

Fig. 8.4 Landscape transformations to and from a land use type, *within* the area of land use types

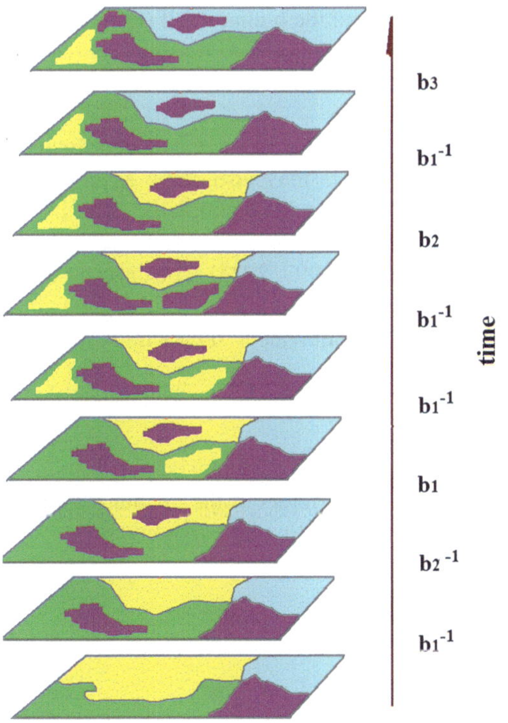

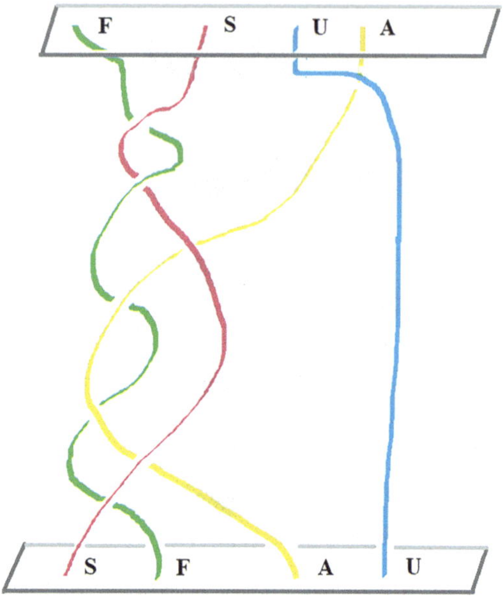

Fig. 8.5 The braid model for he visualization of the landscape changes of Fig. 8.4

strand. Hence, there are two *amphicheiral* representations of the knot 5_1: the a-cinquefoil and the b-cinquefoil (Fig. 8.7).

For the a-cinquefoil the crossings are (clockwise) aI, aII, aIII, aIV, and aV, while for the b-cinquefoil the crossings are counted counter-clockwise: bI, bII, bIII, bIV, bV (Table 8.1).

Besides their obvious usefulness for better visualizations of landscape changes, knots and braids may also serve as *qualitative* models of quantitative land use changes. A formidable example is given by the percentages of land use areas that have been documented for the region of East Attica (Greece), from 1948 until 2022 (Papadimitriou, 1997, 2023). The diagram depicting these changes (Fig. 8.8) can be described qualitatively by the braid of the knot 4_1 (the "figure-of-eight" or "savoy" knot) which is equivalent to the braid word:

$$\sigma_1 \sigma_2{}^{-1} \sigma_1 \sigma_2{}^{-1}$$

In this way, all the characteristic polynomial descriptions of the 4_1 knot apply as quantitative descriptors of the qualitative model of land use change of East Attica; specifically, the following polynomials:

Alexander:

$$-x - \frac{1}{x} + 3$$

Conway:

$$1 - x^2$$

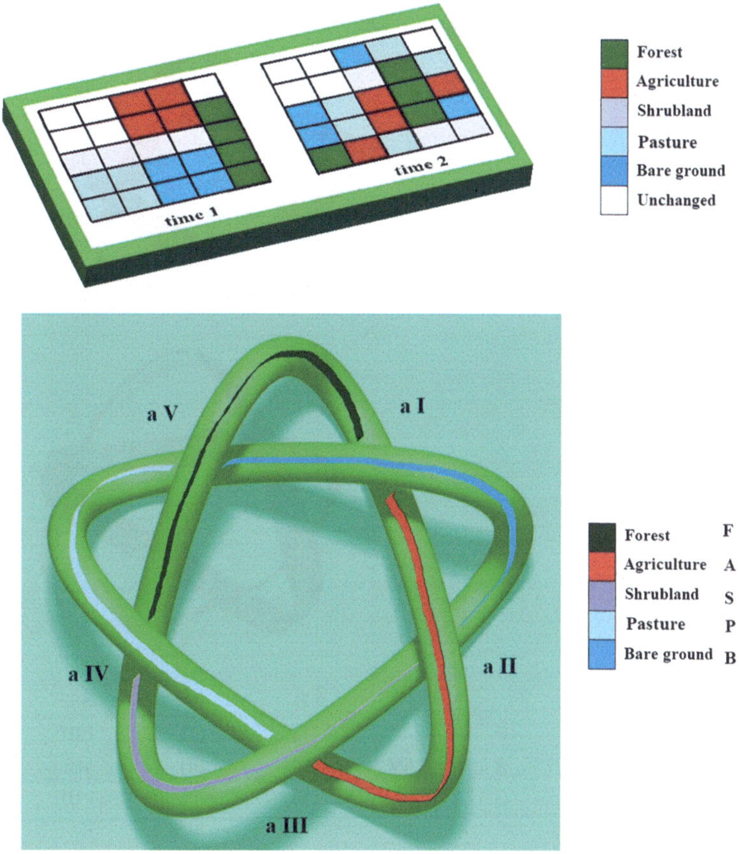

Fig. 8.6 Transformations of an example landscape (**a**) with five land use types (F, A, S, P, B) and areas that remained unchanged (U), together with its cinquefoil model (**b**) illustrating the crossings from aI to aV

Homfly:

$$x^2 + \frac{1}{x^2} - y^2 - 1$$

Jones:

$$x^2 + \frac{1}{x^2} - x - \frac{1}{x} + 1$$

Kauffman:

$$x^2y^2 + \frac{y^2}{x^2} - x^2 - \frac{1}{x^2} + xy^3 + \frac{y^3}{x} - xy - \frac{y}{x} + 2y^2 - 1$$

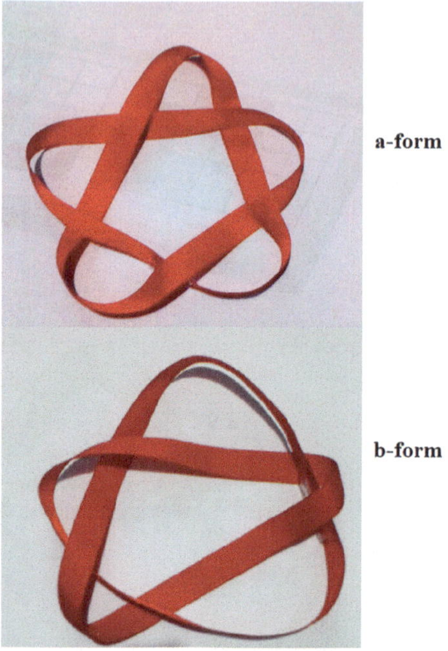

Fig. 8.7 The two amphicheiral forms of the cinquefoil

Table 8.1 Correspondence of the a-cinquefoil and b-cinquefoil crossings to the landscape transformations

	F	B	A	S	P
F	–	bI	aII	bIII	aIV
B	bV	–	bII	aIII	bV
A	bI	bI	–	bIII	aIV
S	aV	bII	aIII	–	bIV
P	bV	aI	bIII	bIII	–

8.3 Knotted Space-Time-Lines

Furthermore, the usefulness of knots in geo-visualization is not confined to modelling land use transformations; knots might also be used to visualize timelines in space. This expands their range of applications to geographical analysis, considering that the visualization and analysis of daily or personal activities is a recurrent theme of research in *time geography* (Chapin Jr., 1974; Carlstein et al., 1978; Adams, 1995; Plaisant et al., 1998; Dykes & Mountain, 2003; Ellegård & Cooper, 2004; Couclelis, 2009; Ellegård & Svedin, 2012; Ellegård, 2018; Dijst, 2020) in which various different visualizations of timelines in the geographical space have been carried out, exploring a variety of types of human movement in space (Kwan, 2004; Kwan & Lee, 2004; Kapler & Wright, 2005; Andrienko & Andrienko, 2006, 2008; Shaw et al., 2008; Andrienko & Andrienko, 2010; Chen et al., 2011; Andrienko & Andrienko 2013; Andrienko et al., 2007, 2010, 2013; Parent et al., 2013; Von

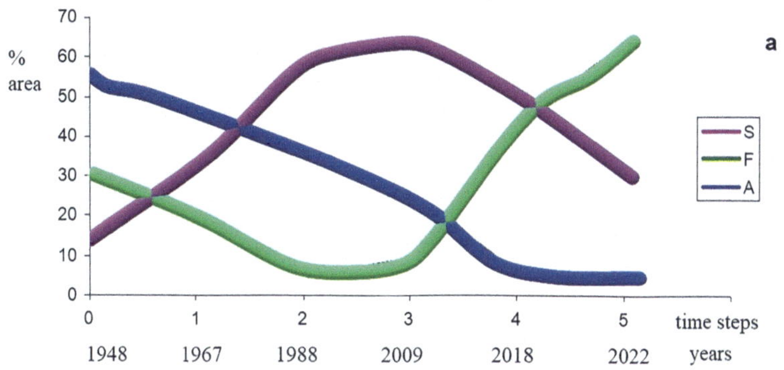

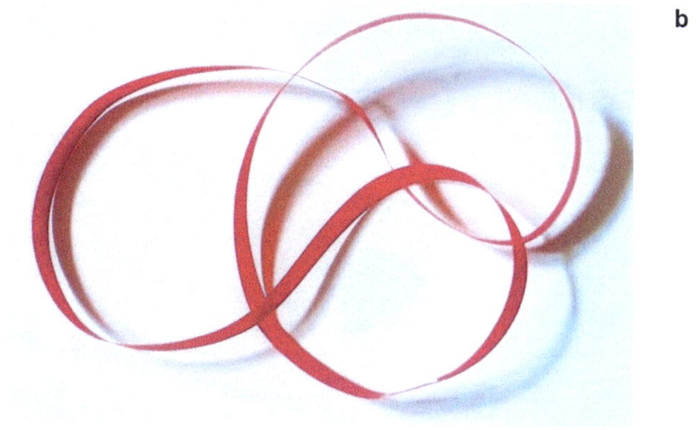

Fig. 8.8 The changes in East Attica, Greece, from 1948 until 2022 (**a**) and the 4_1 knot (**b**)

Landesberger et al., 2015). Nonetheless, these studies mainly concern human movement in open spatial domains, so if the movements take place with a precisely confined region of space and with repeating itineraries and if the end point coincides with the beginning of the movement, then a knot might be an appropriate visualization of such movements in the geographical space (i.e. movements confined in a house, in a park, or any other close spatial domains).

A characteristic example is the timeline of one's stroll in a park: if a graph were used to visualize it, then, upon crossing a junction in the graph, the graph would be inadequate to show whether a left or a right turn was taken. But if a continuous curve were used in 3D space, it would be suitable to represent unambiguously the exact locations that were visited (and with the order they were visited). So in this case knot models provide room for more flexible visualizations than ordinary graphs (Fig. 8.9). Besides, it may also be of interest to landscape architects to design knots in landscaped gardens (Fig. 8.10) for recreational or even for educational purposes.

Fig. 8.9 Visualizing a stroll in a park with the aid of a knot

Fig. 8.10 Using a trefoil knot (and indeed any other knot) in landscape design

References

Adams, P. C. (1995). A reconsideration of personal boundaries in space-time. *Annals of the Association of American Geographers, 85*, 267–285.

Andrienko, N., & Andrienko, G. (2006). *Exploratory analysis of spatial and temporal data: A systematic approach*. Springer.

Andrienko, G., & Andrienko, N. (2008). Spatio-temporal aggregation for visual analysis of movements. In *2008 IEEE symposium on visual analytics science and technology* (pp. 51–58). IEEE.

Andrienko, N., & Andrienko, G. (2010). Spatial generalization and aggregation of massive movement data. *IEEE Transactions on Visualization and Computer Graphics, 17*(2), 205–219.

Andrienko, N., & Andrienko, G. (2013). Visual analytics of movement: An overview of methods, tools and procedures. *Information Visualization, 12*(1), 3–24.

Andrienko, G., Andrienko, N., Jankowski, P., Keim, D., Kraak, M. J., MacEachren, A., & Wrobel, S. (2007). Geovisual analytics for spatial decision support: Setting the research agenda. *International Journal of Geographical Information Science, 21*(8), 839–857.

Andrienko, G., Andrienko, N., Demsar, U., Dransch, D., Dykes, J., Fabrikant, S. I., et al. (2010). Space, time and visual analytics. *International Journal of Geographical Information Science, 24*(10), 1577–1600.

Andrienko, G., Andrienko, N., Bak, P., Keim, D., & Wrobel, S. (2013). *Visual analytics of movement*. Springer.

Artin, E. (1947). Theory of braids. *Annals of Mathematics, 48*, 101–126.

Bangert, P. D., Berger, M. A., & Prandi, R. (2002). In search of minimal random braid configurations. *Journal of Physics A: Mathematical and General, 35*(1), 43–59.

Barenghi, C. F., Ricca, R. L., & Samuels, D. C. (2001). How tangled is a tangle? *Physica D, 157*(3), 197–206.

Berger, M. A. (1994). Minimum crossing numbers for 3-braids. *Journal of Physics A: General Physics, 27*(18), 6205–6213.

Carlstein, T., Parkes, D., & Thrift, N. (1978). *Timing space and spacing time II: Human activity and time geography*. Arnold.

Chapin, F. S., Jr. (1974). *Human activity patterns in the City*. Wiley.

Chen, J., Shaw, S. L., Yu, H., Lu, F., Chai, Y., & Jia, Q. (2011). Exploratory data analysis of activity diary data: A space–time GIS approach. *Journal of Transport Geography, 19*(3), 394–404.

Couclelis, H. (2009). Rethinking time geography in the information age. *Environment and Planning A, 41*(7), 1556–1575.

Crowell, R. H., & Fox, R. H. (1963). *Introduction to knot theory*. Blaisdell.

Diao, Y., & Ernst, C. (1998). The complexity of lattice knots. *Topology and its Applications, 90*(1), 1–9.

Dijst, M. (2020). Time geographic analysis. In *International encyclopedia of human geography* (2nd ed., pp. 271–282). Elsevier.

Dykes, J. A., & Mountain, D. M. (2003). Seeking structure in records of spatio-temporal behaviour: Visualization issues, efforts and applications. *Computational Statistics & Data Analysis, 43*(4), 581–603.

Dynnikov, I., & West, B. (2007). On complexity of braids. *Journal of the European Mathematical Society, 9*(4), 801–840.

Ellegård, K. (2018). *Thinking time geography: Concepts, methods and applications*. Routledge.

Ellegård, K., & Cooper, M. (2004). Complexity in daily life–a 3D-visualization showing activity patterns in their contexts. *Electronic International Journal of Time Use Research, 1*(1), 37–59.

Ellegård, K., & Svedin, U. (2012). Torsten Hägerstrand's time-geography as the cradle of the activity approach in transport geography. *Journal of Transport Geography, 23*, 17–25.

Fiedler, B., & Rocha, C. (1999). Realization of meander permutations by boundary value problems. *Journal of Differential Equations, 156*(2), 282–308.

Foufoula-Georgiou, E., & Sapozhnikov, V. (2001). Scale invariances in the morphology and evolution of braided rivers. *Mathematical Geology, 33*, 273–291.

Garber, D., Kaplan, S., & Teicher, M. (2002). A new algorithm for solving the word problem in braid groups. *Advances in Mathematics, 167*(1), 142–159.

Gordon, C. M. (1978). Some aspects of classical knot theory. *Lecture Notes in Mathematics, 685*, 1–60.

Hamidi-Tehrani, H. (2000). On complexity of the word problem in braid groups and mapping class groups. *Topology and its Applications, 105*(3), 237–259.

Hemion, G. (1992). *The classification of knots and 3–dimensional spaces*. Oxford University Press.

Kapler, T., & Wright, W. (2005). Geotime information visualization. *Information Visualization, 4*(2), 136–146.

Kassel, C., & Turaev, V. (2008). *Braid groups*. Springer.

Kauffman, L. H. (1987). *On Knots. Annals of mathematical studies* (p. 115). Princeton University Press.

Kawauchi, A. (1996). *A survey of knot theory*. Birkhäuser.

Kholodenko, A. L., & Rolfsen, D. P. (1996). Knot complexity and related observables from path integrals for semiflexible polymers. *Journal of Physics A: Mathematical and General, 29*(17), 5677–5691.

Kwan, M. P. (2004). GIS methods in time-geographic research: Geocomputation and geovisualization of human activity patterns. *Geografiska Annaler: Series B, Human Geography, 86*(4), 267–280.

Kwan, M. P., & Lee, J. (2004). Geovisualization of human activity patterns using 3D GIS: A timegeographic approach. In M. F. Goodchild & D. G. Janelle (Eds.), *Spatially integrated social science* (pp. 48–66). Oxford University Press.

Lickorish, W. B. R. (1997). *An introduction to knot theory*. Springer.

Makowsky, J. A., & Marino, J. P. (2003). The parametrized complexity of knot polynomials. *Journal of Computer and System Sciences, 67*(4), 742–756.

Markoff, A. (1945). Foundations of the algebraic theory of braids. *Trudy Mathematical Institute Steklov, 16*, 3–53.

Murasugi, K. (2007). *Knot theory and its applications*. Birkhäuser.

Nechaev, S., & Voituriez, R. (2005). Conformal geoemtry and invariants of 3-strand Brownian Braids. *Nuclear Physics B, 714*(FS), 336–356.

Nechaev, S. K., Grosberg, A. Y., & Vershik, A. M. (1996). Random walks on braid groups: Brownian bridges, complexity and statistics. *Journal of Physics A: Mathematical and General, 29*(10), 2411–2433.

Orlandini, E., Tesi, M. C., & Whittington, S. G. (2005). Entanglement complexity of semiflexible lattice polygons. *Journal of Physics A: Mathematical and General, 38*(47), L795–L800.

Papadimitriou, F. (1997). *Land use modelling, land degradation and land use planning in East Attica, Greece*. Ph. D thesis. University of Oxford.

Papadimitriou, F. (2020). Modelling and visualization of landscape complexity with braid topology. In D. Edler, C. Jenal, & O. Kühne (Eds.), *Modern approaches to the visualization of landscapes* (pp. 79–101). Springer.

Papadimitriou, F. (2023). *Modelling landscape dynamics*. Springer.

Parent, C., Spaccapietra, S., Renso, C., Andrienko, G., Andrienko, N., Bogorny, V., et al. (2013). Semantic trajectories modeling and analysis. *ACM Computing Surveys (CSUR), 45*(4), 1–32.

Piégay, H., Grant, G., Nakamura, F., & Trustrum, N. (2006). Braided river management: From assessment of river behaviour to improved sustainable development. *Braided Rivers: Process, Deposits, Ecology and Management, 36*, 257–275.

Plaisant, C., Shneiderman, B., & Mushlin, R. (1998). An information architecture to support the visualization of personal histories. *Information Processing and Management, 34*(5), 581–597.

Shaw, S. L., Yu, H., & Bombom, L. S. (2008). A space-time GIS approach to exploring large individual-based spatiotemporal datasets. *Transactions in GIS, 12*(4), 425–441.

Simsek, H., Bayram, M., & Can, I. (2003). Automatic calculation of minimum crossing numbers of 3-braids. *Applied Mathematics and Computation, 144*(2–3), 507–516.

Thiffeault, J.-L. (2005). Measuring topological chaos. *Physics Review Letters, 94*, 084502.

Von Landesberger, T., Brodkorb, F., Roskosch, P., Andrienko, N., Andrienko, G., & Kerren, A. (2015). Mobility graphs: Visual analysis of mass mobility dynamics via spatio-temporal graphs and clustering. *IEEE Transactions on Visualization and Computer Graphics, 22*(1), 11–20.

Williams, R. D., Brasington, J., & Hicks, D. M. (2016). Numerical modelling of braided river morphodynamics: Review and future challenges. *Geography Compass, 10*(3), 102–127.

Part IV
Topological Models for Cyber-Geography

Chapter 9
Topologies of Ubiquity and Placelessness

If the path is beautiful, we don't ask where it leads to
"Si le chemin est beau, ne nous demandons pas où il mène"
(Anatole France, 1844–1924, "La vie littéraire")

9.1 Models of Ubiquitous Connectedness

The fast increase in the world's population is associated with a rapid growth in the numbers of users of social media and geospatial technologies; larger numbers of users imply higher average connectivity. Thus, the denser the networks of local, regional or international connections become, the more important the ensuing large-scale topologies. Consider, for instance, the growth in size of an *Erdős–Rényi* network from 100 users to 1000 users: the links increase fast, to the extent that mapping them out means that the whole area in between the nodes would be covered by the links among them (Fig. 9.1). This explosive growth has led to a "hyperconnectivity", one of the hallmarks of modern urban societies (Cheok, 2016; Otrel-Cass, 2019; Brubaker, 2020; Dawson, 2020; Calvo, 2020; Bibri & Allam, 2022; Vermesan et al., 2022). In this new reality, the relationships of the physical space with the virtual space become all the more significant.

The Internet is a network "open" to new nodes (accepting new incoming webpages) and each webpage is neither randomly nor uniformly correlated to any other ones, so it clearly is not a random network. In fact, a random network model would be insufficient to describe the Internet, because it would take into account only closed networks which do not accept new vertices and would assume random connections between any two vertices. What happens instead is that each webpage displays a preferential connectivity to other ones. Most usually, each new webpage incoming to the system is designed to "prefer" an association with specific other webpages.

In fact, exactly as there are preferential connectivities among species (or individuals) in nature and society, so they are in the Internet also. Indeed, according to the "Barabasi and Albert" model which suggests that the probability

© The Author(s), under exclusive license to Springer Nature Switzerland AG 2023
F. Papadimitriou, *Geo-Topology*, GeoJournal Library 133,
https://doi.org/10.1007/978-3-031-48185-7_9

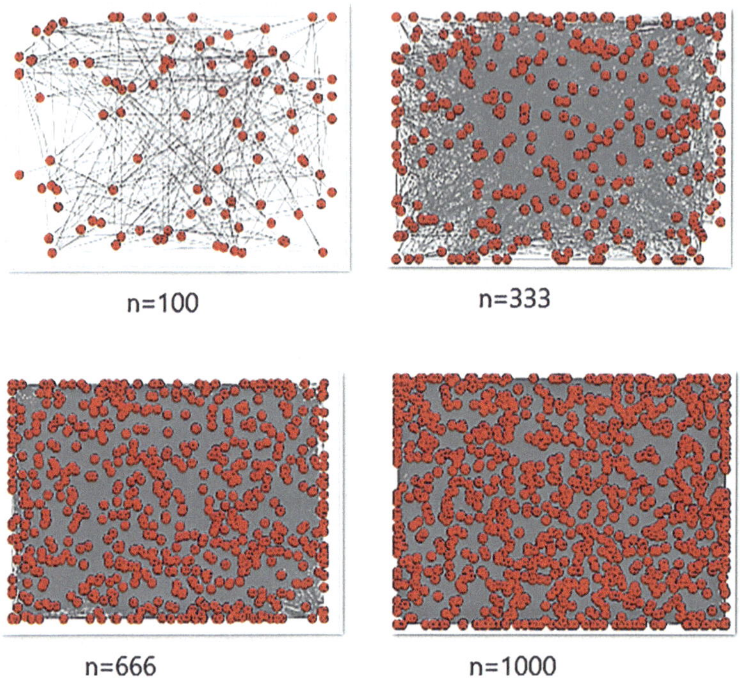

Fig. 9.1 The growth of an ***Erdős–Rényi*** network in terms of nodes results in rapidly increasing numbers of links among nodes also

of a node interacting with (or being linked to) k other nodes displays a scale-invariant distribution and is given by a power law of the type (Barabasi & Albert, 1999):

$$P(k) = k^{-\gamma}$$

Such networks are *scale-free*, since their degree distribution follows a power law for large values of k and deviates significantly from the distribution of nodes N expected from the Poisson formula. It has been claimed (Sole & Valverde, 2001) that the Internet displays signs of self-organisation, but self-organization can not be maintained without some high degree of robustness; the latter being a key property of a modelled system, essentially a measure of the system's capability to survive. While there are "small-world" networks with high circuitry, in the "scale-free network" topology of the Internet some nodes stand out as more important than other ones due to their higher connectivity (Fig. 9.2), so it would be reasonable to assume that the robustness of the Internet can be attributed to its being a "scale-free" network. Some minimum amount of hierarchy is necessary in order to maintain the stability of a scale-free network (social networks may constitute one such form of hierarchy in the Internet). However, it is not precisely known exactly which forms of hierarchic structures should or could prevail in the Internet in order to maintain its stability while also preserving its high complexity.

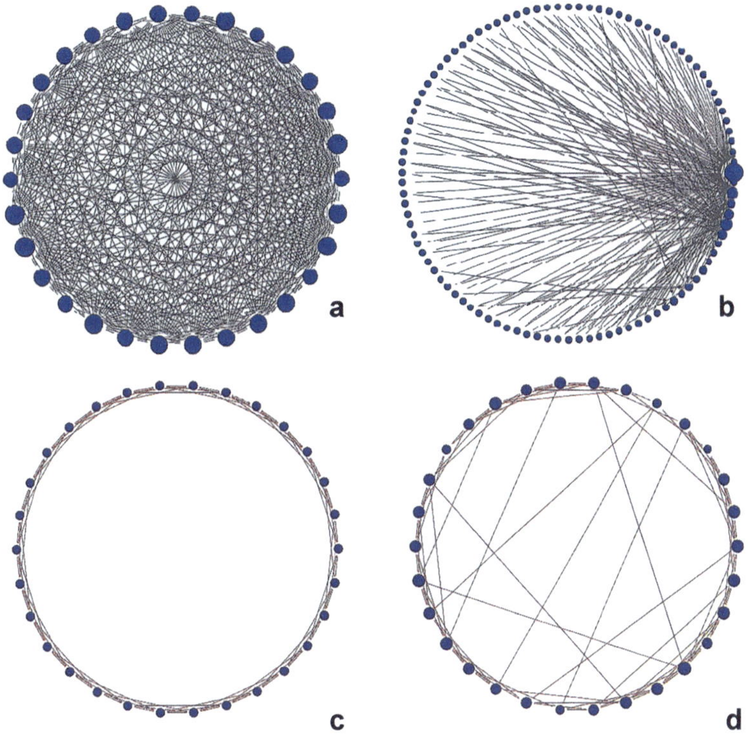

Fig. 9.2 Top: a network with 30 nodes and mean degree per node equal to 30 (**a**) and a "scale-free" network with 100 nodes (**b**). Bottom: two different network topologies with 30 nodes and mean degree per node equal to 6: a regular lattice (**c**) and a "small world" network with a topology that is in between regularity and random connections (**d**)

The topological shape of Hopf links (Fig. 9.3) may be used to visualize the interaction of physical and digital spaces before ubiquitous digital technologies (such as the GPRS, VPNs, WiFi, Bluetooth etc) were introduced, when the digital space was clearly distinguishable from the physical space, when all digital telecommunications were cable-based (e.g. up until the late 1990s) and when the physical and the digital worlds were two non-intersecting worlds (visualized as two non-intersecting tori).

Topologically, the Hopf links are the simplest nontrivial link with at least two components, consisting of two circles linked together exactly once and their braid word is σ_1^2.

Another visualization of non-local interactions in the geographical space which become possible with the use of ICTs can be made by recalling the topological concept of handles. The consideration of handles as a visual metaphor to model non-local connections is possible because handles can serve as tubular models of network links. Nevertheless, handles are not equivalent to 1D links, because links are not hollow from inside (while handles can be so). Further, while the links of a

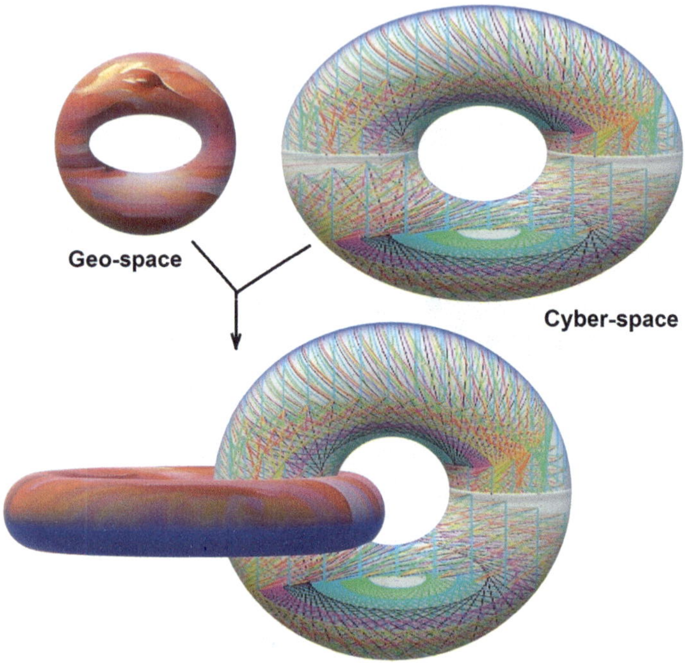

Fig. 9.3 Hopf links from two tori, visualizing the link between the geo-space and the cyber-space when they were separated from one another

Fig. 9.4 Using handles to visualize connections in the geo-space or cyber-space. Handles can have varying degrees of strength and expansion in the cyber-space, e.g.: short distance and weak connection (**a**), short distance and strong connection (**b**), long distance and weak connection (**c**), long distance and strong connection (**d**)

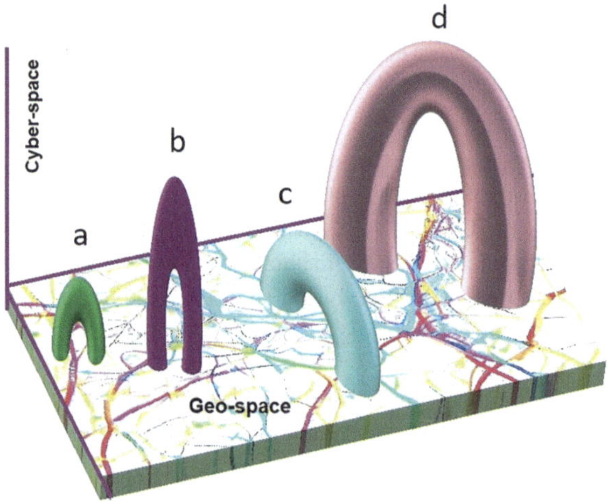

network just connect nodes as lines or curves, handles may connect nodes as geographical areas and may thus model relationships among geographical locations spanning over various distances and with different strengths of connection in the geo-space or in the cyber-space (Fig. 9.4). Plausibly therefore, using a handle to visualize the Internet connections linking the Western Europe with North America should be thicker than that connecting South America with North America.

Currently however, there are several "cyber-ecosystems", such as intranets, VPNs, ISP networks, distributed networks etc. More relevant for geography is that geoinformatics and cartography are not meant for mapping exterior or large geographical areas only; they are being increasingly employed to map small-scale and indoor environments by using "ubiquitous computing" and "ambient intelligence" also, leading to the creation of hybrid physical-and-digital spaces, in which computational intelligence expands over the entire geo-space available to be covered by such technologies. The applications of ambient intelligence and ubiquitous computing emerged in clothing, furniture, toys and many other everyday objects of indoor environments quite some time ago (Weiser, 1991, 1993; Weiser & Brown, 1997; USNRC, 2001; US National Research Council, 2003). Spatially-integrated intelligent environments can be created with these technologies and GIS can be used to build "Smart Homes" (ambient energy harvesting systems, with the aid of wireless sensors implanted or attached to furniture, such as chairs, computers, electric devices, walls etc). Various tasks can be performed with these sensors, i.e. air temperature measurements, monitoring local microclimate, filtering out noise etc., as evidenced from numerous smart home applications (Tapia et al., 2004; Mozer, 2004; Hagras et al., 2004; Youngblood & Cook, 2007; Helal et al., 2005). Whatever the particular technology used, geospatial technologies are indispensable for making hybrid cyber-and-geo-spaces "intelligent". The omnipresence of ICTs with these technologies creates an "instantaneous ubiquity" (Bukatman, 2005, p.120) which may expand widely over urban geo-spaces.

Although not classified as such, geospatial technologies also offer a formidable opportunity to advance ubiquitous connectedness in the geo-cyber-space. The Geospatial Web (Geoweb) entails the Internet as we know it, plus a semantic component, therefore leading to a geographically-enabled Semantic Web. Neogeography emerged along with the Geospatial Web (Haklay et al., 2008; Hudson-Smith et al., 2009; Rana & Joliveau, 2009; Goodchild, 2009; Papadimitriou, 2010a, b; Graham, 2010; Warf & Sui, 2010; Levental, 2012; Lin, 2013; Leszczynski, 2014), and, as its name suggests, is a new geography emerging from the interaction of GIS with GPS, mobile telephony and the Internet, allowing users to "upload" and "download" opinions, views, snapshots, ideas etc. on official or informal maps, resulting in new information structures called "mashups" (Liu & Palen, 2010) that combine images, texts, sounds, videos and symbols on digital maps by allowing users to send images and to "geotag" them (that is to associate coordinates).

The explosive expansion of the WWW in its Post-CERN era (and particularly in the early twenty-first century) saw the development of the Semantic Web and the phenomenon of social networking. Geospatial technologies become all the more important by their interfaces with other technologies (such as GIS and GNSS) which were not initially meant to be compatible with the Internet. Considering the expansion of all these technologies, it perhaps should be expected that, besides physical ecosystems, there is an "Internet ecology" (Mcfedries, 2003), which is populated by "informavores" and "information foragers", constituting "information food chains".

However extreme or metaphorical these expressions may appear to be, the Internet has already been examined as if it were a natural ecosystem, by numerous researchers, for various reasons. Although definitions may vary, the term "Internet of Things Ecosystem" (Mazhelis et al., 2012; Shin & Jin Park, 2017; Leminen et al., 2017; Pradeep et al., 2021; Paolone et al., 2022; Dias et al., 2022) means the complex interplay between varying natural and artificial ecosystems that may include (not restrictively) WiFi, RFID, GPRS, WSAN (wireless sensor and actuator networks), ubiquitous computing, Web of Things (WoT) and other technologies which basically relate biota (humans or plants or other living beings) with technological artifacts, under the umbrella concepts of the Internet of "Things", "Bodies" and "Everything". In this context, the term "Internet of Nature" was introduced (Galle et al., 2019) to denote the unification of "green cities" with "digital cities", in which natural elements of an urban ecosystem are combined with a very wide range of technologies that include networks of sensors, satellite imagery, artificial intelligence, augmented reality, geoinformatics, cloud, robotics, even blockchain and cryptocurrencies. Thus is being created what Bukatman would call a "cybernetic paraspace" (Bukatman, 2005, p.166). Perhaps, nowhere is the merger between geospace and cyber-space more explicit than in the film "The Matrix", in which space is reshaped into spacetime. This unified geo-cyber-space allows one of the film's characters ("Trinity"), to "leap across impossible spaces and be reconfigured from a 3D human figure into a signal on a phone line" (Wood, 2007, p.57).

This fusion of physical with virtual spaces corresponds to a radically different topology than the one that was described by the Hopf links. The topological shape of a "cobordism" is well suited to model these cyber-physical spaces, in which it is difficult to distinguish the strands of the "cyber-" from those of the "geo-". A cobordism between manifolds M and N is a compact manifold W of which the boundary is the disjoint union of M and N, i.e. a cobordism as a single circle M found at the bottom resulting from a pair of disjoint circles (manifold N) atop of it (Fig. 9.5). The "pair of pants" is an example of a more general cobordism: for any two *n*-dimensional manifolds M, M′, the disjoint union is cobordant with the

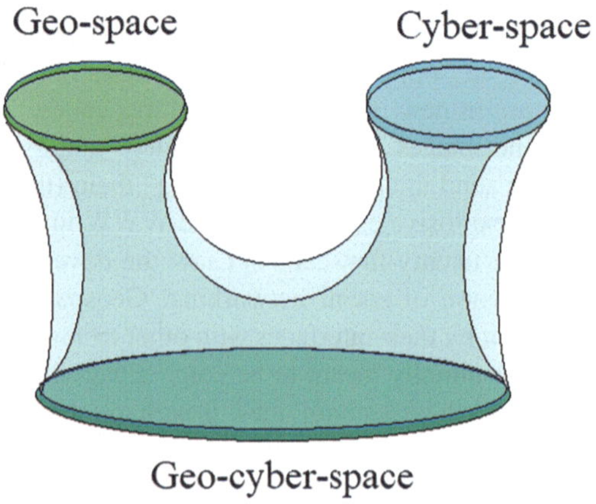

Fig. 9.5 A 3D cobordism as a model of the fusion of geo-space with the cyber-space: a single circle (at the bottom) resulting from the pair of disjoint circles (at the top)

connected sum M#M′ (see Strong, 1968; Milnor, 1962; Pontryagin, 1959; Novikov, 1967; Ravenel, 1986; Atiyah, 1961; Kosinski, 2007).

9.2 Topologies of Placelessness

As known from multiple examples of historical cartography, some representations of the geographical space give a sense of "marvelous" and "edemic" emerging from maps of the past in which cartographers often mapped "approximately" certain unknown regions (Wallace & Van den Heuvel, 2005), or added imaginary elements on such areas, even if such places did not exist in reality. In fact, ever since the concept of placelessness was brought forth and highlighted by Relph (1976), it was explored from within different perspectives, i.e. philosophy (Philippopoulos-Mihalopoulos, 2001), psychology (Arefi, 1999), and prominently geography (Relph, 2000; Hooykaas et al., 2008; Seamon & Sowers, 2008; Freestone & Liu, 2016; Kortelainen & Albrecht, 2021).

Some locations in the real (physical) space have also been considered as "placeless" (Wicomb, 2015) giving the impression that they can just be anyone's place or "no man's land". As "non-places" can be conceived various disparate types of geographical settings, with varying size they can be entire cities such as the soviet "closed cities" Zeleznogorsk, Arzamas-16, Krasnoyarsk-26 etc. that had hosted scientific scientific and military installations (Rowland, 1996; Chen et al., 2016; Glazyrina, 2000; Siddiqi, 2022) and China's "ghost cities" (Yu, 2014; Shepard, 2015; Sorace & Hurst, 2016; Jin et al., 2017; Lu et al., 2018; Woodworth, 2020) as well as unrecognised countries such as the "Principality of Sealand" (Dennis, 2002; Cogliati-Bantz, 2012; Lyon, 2014; Mislan et al., 2019). Yet, they can simply be unclaimed lands, i.e. in the wilderness of Brazil (Holston, 1991), parking lots (Arefi, 2004; Kortelainen & Albrecht, 2021), airports (Rowley & Slack, 1999; Augé, 2020) and hotels (Pritchard & Morgan, 2006). Irrespective of their size, they have also been termed "liminal spaces" (Moore, 2009; Andrews & Roberts, 2012; Ahlrichs et al., 2015; Shortt, 2015; Piazza, 2019).

Placelessness (see also Kunstler, 1994) is a feeling that connects humans with the geographical space (Nagel, 1986), whereas "a large intellectual gap exists between our sense of being actors in the world, of always being in place, and the placelessness that characterizes our attempts to theorize about human actions and events" (Entrikin, 1991, p.7), whereas place can "acquire deep meaning for the adult through the steady accretion of sentiment over the years. Every piece of heirloom furniture, or even a stain on the wall tells a story" (Tuan, 1977, p.3).

However, placelessness is not always a personal feeling or a collective impression of some geographical setting; it may as well be a product of human intellect describing inexistent places ("utopias", "dystopias", "eutopias" etc). A characteristic such "no-place" is in the naturalist novel "After London" (Jefferies, 1885): in an unspecified time of the future, the human population of London was eliminated after a devastating disaster and the region was colonised by plants and reverted into wilderness (whether such settings would qualify as dystopias or a green utopias remains up to anyone's taste to decide). Akin to Jefferies' vision, William Morris published his "News from Nowhere" (Morris, 1890), an imaginative description of life in a pre-industrial culture when, at the peak of the industrial revolution, Morris imagined a world without railways, urban sprawls, industry, in which woodlands and meadows will have replaced the areas where buildings once used to stand and humans would live happily, working minimally and pleasurably in a clearly rural utopia. Similarly, Rudyard Kipling's often-quoted "Jungle Book" (Kipling, 1895) focuses on the various cultural restrictions imposed by the western civilization. The motto reflecting the basic concepts of the utopias of Jefferies, Morris and Kipling would probably be "et in Arcadia ego". But utopias needn't necessarily be non-urban settings only. Ebenezer Howard's utopia for instance is rural-urban utopia (Howard, 1898), a self-sufficient garden-city with precisely defined population (32,000) and area (6000 acres), replete with open spaces, a central park and a greenbelt. In synthesizing rural and urban modes of life, Howard's utopia was a model of coexistence of rural and urban spaces for geographers, planners and architects ever since (and he might be perceived as a "utopist" as much as a "geographer"). But rural utopias did not monopolize utopist thinking: the architect-utopist Le Corbusier, with his "Charter of Athens" and his concept of a "cité radieuse" had a lasting impact on architects, geographers and planners worldwide. Similarly, in Russia, "de-urbanists" dominated the country's geo-utopist thinking at the turn of the nineteenth century and after. In Japan and in the context of the utopist Kisho Kurokawa's theory of "metabolism" (Pernice, 2022) cities can be conceived as living organisms, so utopist plans were laid out on the basis of this approach in the 1960s; a line of thought that was adopted by other Japanese utopists, i.e. Kenzo Tango and Kiyonuri Kikutake. It may not be coincidental that most of these urban utopias were characterized by large spaces hosting cultural events for the masses as well as ample green areas and parks and, in this way, reflecting some of the aspirations encapsulated in the design of modern "eutopias". Whether rural or urban or mixed however, utopias have no place; all they are is imaginary geographies of placelessness, "no-places".

Yet, "no-places" have not been conceived for the physical space only; the cyber-space also hosts "placeless" domains (Gabrielian & Hirsch, 2018; Halstead, 2021). The theory of "notopia" (from no+topos = prohibited place) purported to give a geographical explanation of the hidden cyber-spaces of the Internet (Papadimitriou, 2006a, b, 2009). Notopia is the set of uncharted locations of the Internet that are used by hackers, "hacktivists" (=hackers+activists), "map hackers", cyberwarriors and other "notopians" who have their physical hideaways dispersed in various geographical locations. Also, "notopians" have their hideaways in uncharted areas of the cyber-space and "anonymize", "clear" or destroy their cyber-searches and, upon entering some location of the cyber-space, they seldom stay there for more than a few fractions of a second. By relying on "technologies of disappearance" (Kroker, 2004, p.173), notopia is not the refuge of experienced hackers alone, but of ordinary Internet users also, who intentionally anonymize their cyber-presence and cyber-searches. Like nomadic individuals, notopians avoid settling down and refuse subjection to unwanted digital surveillance, by becoming "electronomadics"

(Mitchell, 2003, p.167). The classification of spatial distributions by Deleuze and Guattari (1987) is appropriate in this context, since there are "nomadic" distributions (stochastic and resistant to hierarchies) and "sedentary" distributions (categorized and structured) in both physical and virtual spaces. In such cyber-places "the individual self can inhabit ever more loci, in ways that are perceived synchronically even by the self, and thus can swiftly weave different lives, which do not necessarily merge" (Floridi, 2002, p.124) and so creating a "horror vacui semantici" (op.cit., p.123). By masking completely their origins and geographical locations, notopians resemble those living in "extraterritorial networks traversed and (in) the nowherevilles" as "cultural hybrids" living "chez soi" (Bauman, 2005,p.29), displaying a "permanently impermanent self, completely incomplete, definitely indefinite – and authentically inauthentic" (op.cit., p.33).

It is certainly not by accident that the use of spatial terms in the examination of the dangerous areas of the cyber-space is exemplified by the extensive use of the word "landscape", i.e. as a constituent of the expression "cyber threat landscape" (Choo, 2011, 2011a; Gil et al., 2014; Reghunadhan, 2018; Pour et al., 2019; Kaloudi & Li, 2020; Zoppi et al., 2021; Chatzis & Stavrou, 2022; Alhajjar & Lee, 2022; Gkoktsis, 2022). By this term is meant the mapping out of threats to the security of a website or an electronic devise that is connected to the Internet and can possibly be susceptible to malicious activity. It is also used to denote comprehensive threat analyses of cyber-security at some geographical scale (i.e. national, continental or global) in which case the geographical distribution of threats resembles the highs and lows of a physical landscape, just like the topography of mountains and valleys of physical landscapes.

When a society becomes complex up to a certain set of thresholds (which are most likely very hard to determine), it may become increasingly fragile also, to the extent that even a small perturbation might make it collapse. In the current state of our civilization, the stability of network infrastructures after disturbance is crucial for its survival. The stability and resilience of networks with high connectivity pose difficult problems, particularly in case these relate to critical infrastructures. Yet, it is of utmost importance to determine the ranges of stability and resilience of such networks. Generally, it is precisely the topology of a network that defines its thresholds of tolerance since, the higher the number of links the higher the difficulty to identify which one of them is crucial for the network's resilience. In the geo-space, research has focused on exploring the adaptability of landscape networks to change (Lenton & van Oijen, 2002; Folke et al., 2005; Nelson et al., 2007; Henning et al., 2021). In the cyber-space, the conditions under which the Internet and the WWW are tolerant to errors and attacks are known (Aiello et al., 2000; Krapivsky et al., 2000; Barabasi et al., 2001; Albert & Barabasi, 2002), ever since Albert and Barabasi (2002) showed that the average path length l on the Internet is unaffected by a random removal of 60% of its nodes, while large connected clusters may persist under random removals. Given that scale-free random networks are more robust than random networks, in the case of the Internet we eventually know that its stability requires scale-free random networks with power-law distributions and a minimum of hierarchical structuring.

A computer virus may cause instability, loss of resilience or even failure of part of the Internet. Similarly, in the geo-space, a virus may incapacitate a sizeable part of the world economy (as Covid19 virus did). Hence, whether an epidemic caused by a biological virus or by a computer virus, becoming infected is always a matter of contact. Given these, a variety of complex topological epidemic models have been proposed (Ganesh et al., 2005; Chen & Chen, 2010; Pinto et al., 2020; Hirata, 2020), exploring epidemic dynamics on complex networks (Pastor-Satorras & Vespignani, 2001; Pastor-Satorras et al., 2001; Tao et al., 2006; Dorogovtsev et al., 2008; Ferreira et al., 2012; Castellano & Pastor-Satorras, 2017; Wang et al., 2017), which can be correlated (Chen et al., 2018), adaptive (Guo et al., 2013), multiplex (Granell et al., 2013), scale-free (Barthélemy et al., 2004) or even networks of malware ecologies (Szor, 2005; Baumgartner, 2007; Crandall et al., 2008). What governs the possibility to contact someone or something either in the physical space or in the cyber-space is nothing else than the topology of the geographical setting.

Eventually, the topology of contacts depends on the links (physical or electronic) or adjacencies in space and time. What portrays the wiring of nodes and their connections is the topological structure of a network that develops from contacts. Star-shaped networks for instance, facilitate contagion and spread an infection faster than elongated random or simply elongated networks. So different network topologies endow a network with different inherent capabilities for resistance and resilience (Fig. 9.6) and, although combinations of different topologies can easily be constructed, some general rules governing the role of topology in a network's resilience can be stated.

In a bus topology, failure in one node does not affect the other ones, because they are all connected to some central artery. Failure of that artery brings about the failure of the entire network.

In a star topology, if the central node fails, the entire network fails too. But if one peripheral node does, the system remains resistant. Also, if one of the links fails, then only the node that corresponds to that link will fail.

In a ring topology, the failure of one node incapacitates all the other ones.

In a tree topology, failure of one node affects the stability of only some branches and hence of part of the entire network only.

The full mesh topology is probably the most reliable of all, in which any damaged link does not affect the stability of the entire network.

Finally, in a partial mesh topology, stability depends very much on the wiring of each particular network.

The property of connectedness however, does not always imply "*path-connectedness*". The latter means that *any* path connecting two points of some space lies entirely in the same space. Consequently, notopia is not a path-connected space of the Internet. A topological space X is "path-connected" if for any pair of points a and b of X, there is a path $p:[0,1] \rightarrow X$ from a to b. There are some pathological cases where the (general) connectedness does not coincide with path-connectedness, such as the "*topologist's sine curve*" that is the graphic representation of the function $y = \sin(1/x)$ which is path-connected but not locally path-connected (Fig. 9.7). The "*topologist's whirlpool*" is the polar form of the sine curve and might be a suitable

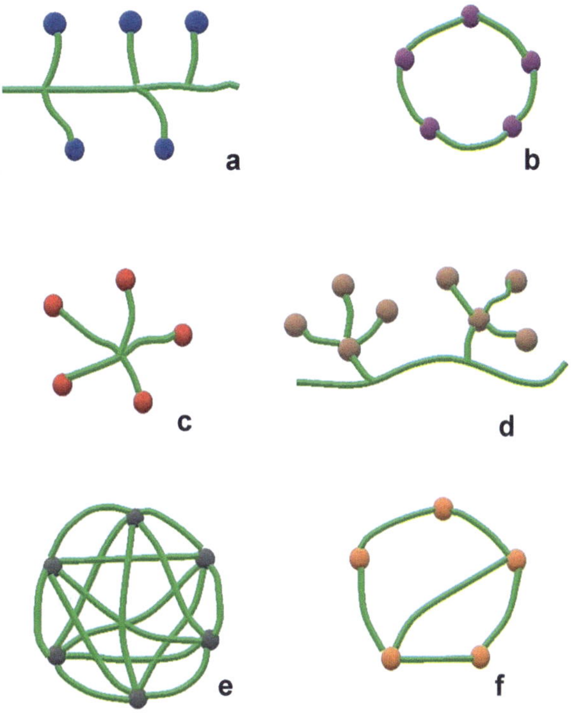

Fig. 9.6 Some characteristic network topologies: bus (**a**), ring (**b**), star (**c**), tree (**d**), full mesh (**e**) and partial mesh (**f**)

topological model for notopia: by concealing all paths to notopia, there is no local path-connection to it from any point in the Internet. The topologist's whirlpool might thus model these virtual no-places, which are connected (by using the communication infrastructures) but not locally path-connected (evidently, this does not apply to utopias, because they are purely imaginary spaces, completely outside the physical reality). In its dynamics, the journey to and from placeless domains in the cyber-space is characteristic of nomad life and, as cyber-nomads, notopians "dwell in disorder and flourish in the midst of dislocation, locating themselves in a network of possibilities" (Sennett, 1998, p.62).

In fact, the cyber-space is replete with "dark" spaces (i.e. the Dark Web) that are inaccessible to most ordinary users, resembling topologically a swiss cheese with holes: a "holey space" in the sense of Deleuze and Guattari (espace troue) populated with the cave-dwellers of the Internet. In the vastness of the "Deep Web" lies an obscure realm that is called "Dark Web" with its cyber-topology of "placelessnes" and, topologically, notopia resembles a *k-connected sphere, or a sphere with k-handles,* where *k* stands for the number of concealed "placeless" IP locations (Fig. 9.8). Notopia is thus the opposite of neogeography: whilst neogeographers want to declare their presence and geographical location in the cyber-space, notopians conceal their presence and hide themselves behind continuously shifting virtual locations, erasing the traces of their actual geographical location. In sharp

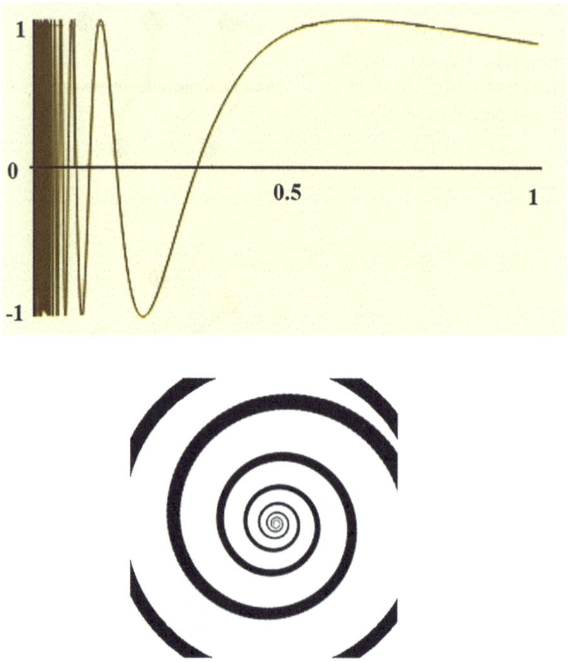

Fig. 9.7 Top: the curve $y = \sin(1/x)$ can serve as an example of a not locally path-connected space. Bottom: Zooming in the "topologist's whirlpool"

Fig. 9.8 A "holey space" with its porus topology of a vast multiply-connected space

contrast to neogeography, in the "personal topologies" of notopia, the prohibited places, the no-go places of the Internet are dispersed all over the world with notopians concealing their presence and hiding themselves behind continuously shifting virtual locations, erasing the traces of their geographical locations. While neogeographers map out their traces in the world, notopians "unmap" theirs. So while neogeographers create an ever denser geo-topology of ubiquity, notopians create a geo-topology of placelessness.

References

Ahlrichs, J. J., Riehle, K., & Sultanalieva, N. (2015). The production of liminal places–an interdisciplinary account. *EAZ–Ethnographisch-Archaeologische Zeitschrift, 56*(1/2), 205–242.

Aiello, W., Chung, F., & Lu, L. (2000). *Proc.32nd ACM symposium on the theory of computing* (p. 171). ACM.

Albert, R., & Barabasi, A.-L. (2002). Statistical mechanics of complex networks. *Reviews of Modern Physics, 74*(1), 47–97.

Alhajjar, E., & Lee, K. (2022). The US cyber threat landscape. *European Conference on Cyber Warfare and Security, 21*(1), 18–24.

Andrews, H., & Roberts, L. (Eds.). (2012). *Liminal landscapes: Travel, experience and spaces in-between*. Routledge.

Arefi, M. (1999). Non-place and placelessness as narratives of loss: Rethinking the notion of place. *Journal of Urban Design, 4*(2), 179–193.

Arefi, M. (2004). The pedagogy of the American city: Revisiting the concepts of place, non-place, and placelessness. *Urban Design International, 9*, 103–117.

Atiyah, M. F. (1961). Bordism and cobordism. *Proceedings of the Cambridge Philosophical Society, 57*, 200–208.

Augé, M. (2020). *Non-places: An introduction to supermodernity*. Verso.

Barabasi, A.-L., & Albert, R. (1999). Emergence of scaling in random networks. *Science, 286*, 509–512.

Barabasi, A. L., Ravasz, E., & Vicsek, T. (2001). Deterministic scale free networks. *Physica A: Statistical Mechanics and Applications, 299*(3–4), 559–564.

Barthélemy, M., Barrat, A., Pastor-Satorras, R., & Vespignani, A. (2004). Velocity and hierarchical spread of epidemic outbreaks in scale-free networks. *Physical Review Letters, 92*(17), 178701.

Bauman, Z. (2005). *Liquid Life*. Polity.

Baumgartner, K. (2007). Storm 2007: Malware 2.0 has arrived. *Virus bulletin*, 2007.

Bibri, S. E., & Allam, Z. (2022). The Metaverse as a virtual form of data-driven smart cities: The ethics of the hyper-connectivity, datafication, algorithmization, and platformization of urban society. *Computational Urban Science, 2*(1), 22.

Brubaker, R. (2020). Digital hyperconnectivity and the self. *Theory and Society, 49*, 771–801.

Bukatman, S. (2005). *Terminal identity: The virtual subject in postmodern science fiction*. Duke University Press.

Calvo, P. (2020). The ethics of Smart City (EoSC): Moral implications of hyperconnectivity, algorithmization and the datafication of urban digital society. *Ethics and Information Technology, 22*(2), 141–149.

Castellano, C., & Pastor-Satorras, R. (2017). Relating topological determinants of complex networks to their spectral properties: Structural and dynamical effects. *Physical Review X, 7*(4), 041024.

Chatzis, P., & Stavrou, E. (2022). Cyber-threat landscape of border control infrastructures. *International Journal of Critical Infrastructure Protection, 36*, 100503.

Chen, P. Y., & Chen, K. C. (2010). Information epidemics in complex networks with opportunistic links and dynamic topology. In *2010 IEEE global telecommunications conference GLOBECOM 2010* (pp. 1–6). IEEE.

Chen, M., Liu, W., & Lu, D. (2016). Challenges and the way forward in China's new-type urbanization. *Land Use Policy, 55*, 334–339.

Chen, X. H., Cai, S. M., Wang, W., Tang, M., & Stanley, H. E. (2018). Predicting epidemic threshold of correlated networks: A comparison of methods. *Physica A: Statistical Mechanics and its Applications, 505*, 500–511.

Cheok, A. D. (2016). *Hyperconnectivity*. Springer.

Choo, K. K. R. (2011). The cyber threat landscape: Challenges and future research directions. *Computers & Security, 30*(8), 719–731.

Choo, K. K. R. (2011a). Cyber threat landscape faced by financial and insurance industry. *Trends and Issues in Crime and Criminal Justice [electronic resource], 408*, 1–6.
Cogliati-Bantz, V. (2012). My platform, my state: The principality of Sealand in international law. *Journal of International Maritime Law, 18*(3), 227–250.
Crandall, J. R., Ladau, J., Ensafi, R., Shebaro, B., & Forrest, S. (2008). *The ecology of malware, proceedings of the new security paradigms workshop (NSPW '08)* (pp. 99–106). Lake Tahoe.
Dawson, M. (2020). Exploring secure computing for the internet of things, internet of everything, web of things, and hyperconnectivity. In *Securing the internet of things: Concepts, methodologies, tools, and applications* (pp. 1186–1195). IGI Global.
Deleuze, G., & Guattari, F. (1987). *A thousand plateaus*. University of Minnesota Press.
Dennis, T. A. (2002). The principality of Sealand: Nation building by individuals. *Tulsa Journal of Comparative and International Law, 10*(1), 261–296.
Dias, J. P., Restivo, A., & Ferreira, H. S. (2022). Designing and constructing internet-of-things systems: An overview of the ecosystem. *Internet of Things, 19*, 100529.
Dorogovtsev, S. N., Goltsev, A. V., & Mendes, J. F. (2008). Critical phenomena in complex networks. *Reviews of Modern Physics, 80*(4), 1275.
Entrikin, J. N. (1991). *The Betweenness of place*. Macmillan.
Ferreira, S. C., Castellano, C., & Pastor-Satorras, R. (2012). Epidemic thresholds of the susceptible-infected-susceptible model on networks: A comparison of numerical and theoretical results. *Physical Review E, 86*(4), 041125.
Floridi, L. (2002). What is the philosophy of information? *Metaphilosophy, 33*(1–2), 123–145.
Folke, C., Hahn, T., Olsson, P., & Norberg, J. (2005). Adaptive governance of social-ecological systems. *Annual Reviews Environment Resources, 30*, 441–473.
Freestone, R., & Liu, E. (Eds.). (2016). *Place and placelessness revisited*. Routledge.
Gabrielian, A., & Hirsch, A. B. (2018). Prosthetic landscapes: Place and Placelessness in the digitization of memorials. *Future Anterior: Journal of Historic Preservation, History, Theory, and Criticism, 15*(2), 113–131.
Galle, N. J., Nitoslawski, S. A., & Pilla, F. (2019). The internet of nature: How taking nature online can shape urban ecosystems. *The Anthropocene Review, 6*(3), 279–287.
Ganesh, A., Massoulié, L., & Towsley, D. (2005). The effect of network topology on the spread of epidemics. In *Proceedings IEEE 24th annual joint conference of the IEEE computer and communications societies* (Vol. 2, pp. 1455–1466). IEEE.
Gil, S., Kott, A., & Barabási, A. L. (2014). The cyber threat landscape: Challenges and future research directions. *Scientific Reports, 4*, 5659.
Gkoktsis, G. (2022). Assessing the cyber threat landscape for virtual power plants. *Latin-American Journal of Computing, 9*(2), 22–35.
Glazyrina, V. (2000). Krasnoiarsk-26: A closed city of the defence-industry complex. In *The soviet Defence-industry complex from Stalin to Khrushchev* (pp. 195–202). Palgrave Macmillan UK.
Goodchild, M. (2009). 6th International conference on mining science and technology. *Procedia Earth and Planetary Science, 1*, 1037–1043.
Graham, M. (2010). Neogeography and the palimpsests of place: Web 2.0 and the construction of a virtual earth. *Tijdschrift voor Economische en Sociale Geografie, 101*(4), 422–436.
Granell, C., Gómez, S., & Arenas, A. (2013). Dynamical interplay between awareness and epidemic spreading in multiplex networks. *Physical Review Letters, 111*(12), 128701.
Guo, D., Trajanovski, S., van de Bovenkamp, R., Wang, H., & Van Mieghem, P. (2013). Epidemic threshold and topological structure of susceptible-infectious-susceptible epidemics in adaptive networks. *Physical Review E, 88*(4), 042802.
Hagras, H., Callaghan, V., Colley, M., Clarke, G., Pounds-Cornish, A., & Duman, H. (2004). Creating an ambient-intelligence environment using embedded agents. *Intelligent Systems, IEEE., 19*(6), 12–20.
Haklay, M., Singleton, A., & Parker, C. (2008). Web mapping 2.0: The neogeography of the GeoWeb. *Geography Compass, 2*(6), 2011–2039.

Halstead, H. (2021). Cyberplace: From fantasies of placelessness to connective emplacement. *Memory Studies, 14*(3), 561–571.

Helal, S., Mann, W., El-Zabadani, H., King, J., Kaddoura, Y., & Jansen, E. (2005). The gator tech smart house: A programmable pervasive space. *Computer, 38*(3), 50–60.

Henning, M., Herrmann, P., Zimmermann, T., Meier, M., Pietsch, M., & Schmidt, C. (2021). A scenario and monitoring based planning approach to strengthen the resilience of the cultural landscape. *Journal of Digital Landscape Architecture, 6-2021*, 123–132.

Hirata, Y. (2020). Topological epidemic model: Theoretical insight into underlying networks. *Chaos: An Interdisciplinary Journal of Nonlinear Science, 30*(10), 101103.

Holston, J. (1991). The misrule of law: Land and usurpation in Brazil. *Comparative Studies in Society and History, 33*(4), 695–725.

Hooykaas, A. L., Beer, M. S., & Faizi, M. (2008). The study of placelessness: Toward a conceptual framework. *International Journal of Environmental, Cultural, Economic and Social Sustainability, 5*(3), 295–304.

Howard, E. (1898/2010). *Tomorrow: A peaceful path to reform.* Cambridge University Press.

Hudson-Smith, A., Crooks, A., Gibin, M., Milton, R., & Batty, M. (2009). NeoGeography and Web 2.0: Concepts, tools and applications. *Journal of Location Based Services, 3*(2), 118–145.

Jefferies, R. (1885/2017). *After London; or wild England.* Edinburgh University Press.

Jin, X., Long, Y., Sun, W., Lu, Y., Yang, X., & Tang, J. (2017). Evaluating cities' vitality and identifying ghost cities in China with emerging geographical data. *Cities, 63*, 98–109.

Kaloudi, N., & Li, J. (2020). The ai-based cyber threat landscape: A survey. *ACM Computing Surveys (CSUR), 53*(1), 1–34.

Kipling, R. (1895/2010). *The jungle book.* Harper Collins.

Kortelainen, J., & Albrecht, M. (2021). Placelessness of urban design and industrial branding in small town planning. *Journal of Urban Design, 26*(4), 405–421.

Kosinski, A. (2007). *Differential manifolds*. Dover Publications.

Krapivsky, P. L., Redner, S., & Leyvraz, F. (2000). *Physics Review Letters, 85*, 4629.

Kroker, A. (2004). *The will to Technology & the Culture of nihilism: Heidegger, Nietzche, & Marx.* University of Toronto Press.

Kunstler, J. H. (1994). *The geography of nowhere.* Simon and Schuster.

Leminen, S., Rajahonka, M., & Westerlund, M. (2017). Actors in the emerging internet of things ecosystems. *International Journal of E-Services and Mobile Applications (IJESMA), 9*(1), 57–75.

Lenton, T. M., & van Oijen, M. (2002). Gaia as a complex adaptive system. *Philosophical transactions of the Royal Society of London. Series B: Biological Sciences, 357*(1421), 683–695.

Leszczynski, A. (2014). On the neo in Neogeography. *Annals of the Association of American Geographers, 104*(1), 60–79.

Levental, S. (2012). A new geospatial services framework: How disaster preparedness efforts should integrate neogeography. *Journal of Map & Geography Libraries, 8*(2), 134–162.

Lin, W. (2013). Situating performative neogeography: Tracing, mapping, and performing "Everyone's East Lake". *Environment and Planning A, 45*(1), 37–54.

Liu, S., & Palen, L. (2010). The new cartographers: Crisis map mashups and the emergence of Neogeographic practice. *Cartography and Geographic Information Science, 37*(1), 69–90.

Lu, H., Zhang, C., Liu, G., Ye, X., & Miao, C. (2018). Mapping China's ghost cities through the combination of nighttime satellite data and daytime satellite data. *Remote Sensing, 10*(7), 1037.

Lyon, A. H. (2014). The principality of Sealand, and its case for sovereign recognition. *Emory International Law Review, 29*(3), 637–671.

Mazhelis, O., Luoma, E., & Warma, H. (2012). Defining an internet-of-things ecosystem. In *Conference on internet of things and smart spaces* (pp. 1–14). Springer.

Mcfedries, P. (2003, April). *The internet ecology*. IEEE Spectrum.

Milnor, J. (1962). A survey of cobordism theory. *L'Enseignement Mathématique. Revue Internationale. IIe Série, 8*, 16–23.

Mislan, D. B., Streich, P., Mislan, D. B., & Streich, P. (2019). To the sea! Sealand and other wannabe states. In I. R. Weird (Ed.), *Deviant cases in international relations* (pp. 15–28). Palgrave.
Mitchell, W. J. (2003). *Me++. The cyborg self and the Networked City*. MIT Press.
Moore, C. (2009). Liminal places and spaces: Public/private considerations. In *Production studies* (pp. 133–147). Routledge.
Morris, W. (1890/1994). *News from nowhere*. Penguin.
Mozer, M. C. (2004). Lessons from an adaptive house. In D. Cook & R. Das (Eds.), *Smart environments: Technologies, protocols and applications* (pp. 273–294).
Nagel, T. (1986). *The view from nowhere*. Oxford University Press.
Nelson, D. R., Adger, W. N., & Brown, K. (2007). Adaptation to environmental change: Contributions of a resilience framework. *Annual Review of Environment and Resources, 32*(1), 395–419.
Novikov, S. (1967). Methods of algebraic topology from the point of view of cobordism theory. *Izvestia Akademii Nauk SSSR Ser. Mat., 31*, 855–951.
Otrel-Cass, K. (2019). Hyperconnectivity and digital reality: An introduction. In *Hyperconnectivity and digital reality: Towards the Eutopia of being human* (pp. 1–8). Springer Nature.
Paolone, G., Iachetti, D., Paesani, R., Pilotti, F., Marinelli, M., & Di Felice, P. (2022). A holistic overview of the internet of things ecosystem. *IoT, 3*(4), 398–434.
Papadimitriou, F. (2006a). A geography of Notopia: Hackers et al, hacktivists, urban cyber-groups/cyber-cultures and digital social movements. *City: Analysis of Urban Trends, Culture, Theory, Policy and Action, 10*(3), 317–326.
Papadimitriou, F. (2006b). The cultures of cyberspace. In J. Lidstone (Ed.), *Cultural issues of our time* (pp. 178–187). Cambridge University Press.
Papadimitriou, F. (2009). A nexus of cyber-geography and cyber-psychology: Topos/Notopia and identity in hacking. *Computers in Human Behavior, 25*, 1331–1334.
Papadimitriou, F. (2010a). Introduction to the complex geospatial web in geographical education. *International Research in Geographical and Environmental Education, 19*(1), 53–56.
Papadimitriou, F. (2010b). A "Neogeographical education"? The geospatial web, GIS and digital art in adult education. *International Research in Geographical and Environmental Education, 19*(1), 71–74.
Pastor-Satorras, R., & Vespignani, A. (2001). Epidemic dynamics and endemic states in complex networks. *Physical Review E, 63*(6), 066117.
Pastor-Satorras, R., Vázquez, A., & Vespignani, A. (2001). Dynamical and correlation properties of the internet. *Physical Review Letters, 87*(25), 258701.
Pernice, R. (2022). *The urbanism of metabolism visions, scenarios and models for the Mutant City of tomorrow*. Taylor & Francis.
Philippopoulos-Mihalopoulos, A. (2001). Mapping utopias: A voyage to placelessness. *Law and Critique, 12*, 135–157.
Piazza, R. (Ed.). (2019). *Discourses of identity in liminal places and spaces*. Routledge.
Pinto, E. R., Nepomuceno, E. G., & Campanharo, A. S. (2020). Impact of network topology on the spread of infectious diseases. *TEMA (São Carlos), 21*, 95–115.
Pontryagin, L. (1959). Smooth manifolds and their applications in homotopy theory. *American Mathematical Society Translations, Ser.2, 11*, 1–114.
Pour, M. S., Bou-Harb, E., Varma, K., Neshenko, N., Pados, D. A., & Choo, K. K. R. (2019). Comprehending the IoT cyber threat landscape: A data dimensionality reduction technique to infer and characterize internet-scale IoT probing campaigns. *Digital Investigation, 28*, S40–S49.
Pradeep, P., Krishnamoorthy, S., Pathinarupothi, R. K., & Vasilakos, A. V. (2021). Leveraging context-awareness for internet of things ecosystem: Representation, organization, and management of context. *Computer Communications, 177*, 33–50.
Pritchard, A., & Morgan, N. (2006). Hotel Babylon? Exploring hotels as liminal sites of transition and transgression. *Tourism Management, 27*(5), 762–772.

Rana, S., & Joliveau, T. (2009). Neogeography: An extension of mainstream geography for everyone made by everyone? *Journal of Location Based Services, 3*, 75–81.

Ravenel, D. C. (1986). *Complex cobordism and stable homotopy groups of spheres*. Academic.

Reghunadhan, R. (2018). Cyber threat landscape of digital India: A critical perspective. *Journal of Polity and Society, 10*(1&2), 37–50.

Relph, E. (1976). *Place and placelessness* (Vol. 67). Pion.

Relph, E. (2000). Classics in human geography revisited, place and placelessness. *Progress in Human Geography, 24*(4), 613.

Rowland, R. H. (1996). Russia's secret cities. *Post-Soviet Geography and Economics, 37*(7), 426–462.

Rowley, J., & Slack, F. (1999). The retail experience in airport departure lounges: Reaching for timelessness and placelessness. *International Marketing Review, 16*(4/5), 363–376.

Seamon, D., & Sowers, J. (2008). Place and placelessness, Edward Relph. *Key Texts in Human Geography, 43*, 51.

Sennett, R. (1998). *The corrosion of character*. W.W. Norton.

Shepard, W. (2015). *Ghost cities of China: The story of cities without people in the world's most populated country*. Bloomsbury.

Shin, D. H., & Jin Park, Y. (2017). Understanding the internet of things ecosystem: Multi-level analysis of users, society, and ecology. *Digital Policy, Regulation and Governance, 19*(1), 77–100.

Shortt, H. (2015). Liminality, space and the importance of 'transitory dwelling places' at work. *Human Relations, 68*(4), 633–658.

Siddiqi, A. (2022). Atomized urbanism: Secrecy and security from the gulag to the soviet closed cities. *Urban History, 49*(1), 190–210.

Sole, R. V., & Valverde, S. (2001). Information transfer and phase transitions in a model of internet traffic. *Physica A, 289*(3), 595–605.

Sorace, C., & Hurst, W. (2016). China's phantom urbanisation and the pathology of ghost cities. *Journal of Contemporary Asia, 46*(2), 304–322.

Strong, R. E. (1968). *Notes on cobordism theory*. Princeton University Press.

Szor, P. (2005). *The art of computer virus research and defense*. Symantec Press.

Tao, Z., Zhongqian, F., & Binghong, W. (2006). Epidemic dynamics on complex networks. *Progress in Natural Science, 16*(5), 452–457.

Tapia, E. M., Intille, S. S., & Larson, K. (2004). Activity recognition in the home using simple and ubiquitous sensors. In *Pervasive* (pp. 158–175). Springer.

Tuan, Y.-F. (1977). *Space and place: The perspective of experience*. University of Minnesota Press.

US National Research Council. (2003). *IT roadmap to a geospatial future*. National Academy Press.

USNRC. (2001). *Embedded, everywhere: A research agenda for networked systems of embedded computers*. National Academy Press.

Vermesan, O., Eisen Hauer, M., Serrano, M., Guillemin, P., Sundmaeker, H., Tragos, E. Z., et al. (2022). The next generation internet of things–hyperconnectivity and embedded intelligence at the edge. In *Next Generation Internet of Things–Distributed Intelligence at the Edge and Human-Machine Interactions* (pp. 19–102). River Publishers.

Wallace, T., & Van den Heuvel, C. (2005). Truth and accountability in geographic and historical visualizations. *The Cartographic Journal, 42*(2), 173 181.

Wang, W., Tang, M., Stanley, H. E., & Braunstein, L. A. (2017). Unification of theoretical approaches for epidemic spreading on complex networks. *Reports on Progress in Physics, 80*(3), 036603.

Warf, B., & Sui, D. (2010). From GIS to neogeography: Ontological implications and theories of truth. *Annals of GIS, 16*(4), 197–209.

Weiser, M. (1991). *The computer for the twenty-first century* (pp. 94–10). Scientific American.

Weiser, M. (1993). Some computer science issues in ubiquitous computing. *Communications of the ACM, 36*(7), 75–85.

Weiser, M., & Brown, J. S. (1997). The coming age of calm technology. In P. J. Denning & R. M. Metcalfe (Eds.), *Beyond calculation: The next fifty years of computing*. Copernicus.

Wicomb, Z. (2015). Heterotopia and placelessness in Brian Chikwava's Harare north. In *The globalization of space* (pp. 50–64). Routledge.

Wood, A. (2007). *Digital encounters*. Routledge.

Woodworth, M. D. (2020). Picturing urban China in ruin: "Ghost City" photography and speculative urbanization. *Geo Humanities, 6*(2), 233–251.

Youngblood, G. M., & Cook, D. J. (2007). Data mining for hierarchical model creation. *IEEE Transactions on Systems, Man, and Cybernetics, Part C., 37*(4), 561–572.

Yu, H. (2014). China's "ghost cities". *East Asian Policy, 6*(02), 33–43.

Zoppi, T., Ceccarelli, A., Capecchi, T., & Bondavalli, A. (2021). Unsupervised anomaly detectors to detect intrusions in the current threat landscape. *ACM/IMS Transactions on Data Science, 2*(2), 1–26.

Chapter 10
Ultrametri-City

Love is flower-like; Friendship is a sheltering tree
(Samuel Taylor Coleridge, 1772–1834, "Youth and Age")

10.1 Geo-and-Cyber-Presentity

Networks of living beings constitute the physical, informational and energetic backbone of ecosystems and social systems, as well as of the artificial "ecosystems" that humans have developed in the form of computer and telecommunication networks. Despite the widely held view that the Internet is "somewhere out of here", IP addresses may be linked to precise locations in the geographical (physical) space, the "geo-space". It is thus possible to establish correspondences among flows of information on the Internet and places or regions in the geographical space. In order to establish such correspondences however, it is necessary to recall how a location is defined in both the physical space and the cyber-space.

With the technologies currently available, locations are recorded with high precision with the aid of Global Navigation Satellite Systems (GNSS), such as the American GPS (Global Positioning System), the Russian "Glonass", the Chinese "Baidu" and the European "Galileo". Furthermore, this possibility is also available to those holding GPS receivers or GPS-enabled mobile phone devises.

Identifying a location in the 3D Euclidean space on the surface of the earth by means of a GNSS is mathematically simple, as it takes into account three coordinates only: the directions of the north (n), the east (e) and the vertical to the point of one's location (u). With φ the geographical latitude, λ longitude, R the radius and h the altitude from the surface of the earth, these variables relate to one another by the matrix:

$$\begin{bmatrix} n \\ e \\ u \end{bmatrix} = R \begin{bmatrix} (\mathrm{N}+h)\cos\varphi\cos\lambda \\ (\mathrm{N}+h)\cos\varphi\cos\lambda \\ \left(\mathrm{N}\frac{b^2}{a^2}+h\right)\sin\varphi \end{bmatrix}$$

© The Author(s), under exclusive license to Springer Nature Switzerland AG 2023

F. Papadimitriou, *Geo-Topology*, GeoJournal Library 133,
https://doi.org/10.1007/978-3-031-48185-7_10

whereas the parameter N is defined as:

$$N = \frac{\alpha^2}{\sqrt{\alpha^2 \cos^2\varphi + b^2 \sin^2\varphi}}$$

with α and b the longer and the shorter half-axes respectively of the reference ellipsoid used for the calculation. The American GPS uses the world geodetic standard WGS-84 as a reference system, with half-axes $\alpha = 6{,}378{,}137$ m and $b = 6{,}356{,}752$ m, while other GNSS may use other systems (the Russian GLONASS for instance, uses PZ-90 ECEF).

"Locating" in the cyber-space may accept manifold interpretations. For instance, in Cubitt's (1998, p.88) "Digital Aesthetics", locating oneself on the interface between human beings and cyber-space is equivalent to the position of the cursor on the screen. In fact, there is no metric to measure distances geometrically in the cyber-space, but only topologically so (distance in terms of links and connectance) and technically, an IP address does not immediately correspond to an address in the geo-space, because IP addresses represent locations on the Internet (locations at a world-wide network, to be precise). A computer can be located somewhere in that network but, simultaneously, it can be distant from other computers that share that location. For instance, an IP address might be located at two different geographical sites (e.g. through a remote LAN connection). But the very existence of the Internet is due to the enormously large set of servers located at specific places in the geo-space and this is why the confusion of IP locations with locations in the geo-space is technically possible (that is the "spoofing" technique).

However, the presence of an individual in the cyber-space can be recorded by the PIDF (Presence Identification Format), a document format promoted by the IETF (Internet Engineering Task Force) and written in XML. A PIDF record relates to the GEOPRIV (Geographic Location Privacy working group) of the IETF, which uses a "*presentity target*". The basic PIDF format translates the geodetic location (φ,λ, altitude) to an XML format and thus makes the identification of physical location "understandable" by the Internet; a process further facilitated by the rapidly expanding networks of geosensors (Nittel et al., 2004; Reis, 2005; Lee et al., 2012).

For instance, for a user located in Athens, Greece, at 37.58′N and 23.46′E and at an altitude of 50 m above sea level, the PIDF encoding written in GML (Geography Markup Language, a complex XML structure, specifically devised for handling geodata) is:

```
<gp:location-info>
  <gml:Point srsName="urn:ogc:def:crs:EPSG: :2879">
    <gml:pos>37.58 23.46 50</gml:pos>
  </gml :Point >
</gml :location-info>
```

Further, instead of locations defined by points only, the same coding can be designed so as to handle areas also, i.e. circles around a point:

```
<gp:location-info>
  <gs:Circle srsName="urn:ogc:def:crs:EPSG: :4000">
    <gml:pos>37.58 23.46 </gml:pos>
    <gml:radius uom="urn:ogc:def:uom:EPSG: :7000">
      3322
  </gml:radius >
 </gs:Circle>
</gml:location-info>
```

Besides, localization is also feasible by using RFID tags (transponders of radio frequency identification) that might communicate with RFID readers via Ethernet or any other telecommunication technology. Equally feasible is the localization by means of WLANs (wireless local area networks) with the IEEE 802.11 standard or other equivalent variants that allow wireless nodes to exchange location data and, in this way, a device may acquire new IP addresses as it moves in the geo-space.

10.2 Ultrametric Topologies of Geo-cyber-spaces

Locations in the cyber-space and in the geo-space can be combined together within a unified topology, which can be ultrametric. Before proceeding to defining this topology however, some basic notions of metric spaces should be introduced and the first concept to consider is metrizability.

Metric spaces are the commonest type of topological spaces and their main characteristic is that they are endowed with a means for measuring distances between points in them. A metric space is defined from a set Y and its "metric" function.

A *metric* on a set Y is a function $d{:}Y \times Y$, which endows any three sets of Y with three properties:

(i) $d(a,b) > 0$ for all $a,b \in Y$ and $d(a,b) = 0$ if and only if $a = b$.
(ii) $d(a,b) = d(b, a)$ for all $a,b \in Y$.
(iii) $d(a,b) + d(b,c) \geq d(a,c)$ for all $a,b,c \in Y$ (triangle inequality).

The function $d(a,b)$ is the distance between points a and b.

The distance between the points a and b of the 1D space R is $d(a,b)$, that is the "Euclidean metric" on R. In the same way, the Cartesian distance between two points $a = (a_1,a_2)$ and $b = (b_1,b_2)$ on the plain R^2 is:

$$d(a,b) = \sqrt{(a_1 - b_1)^2 + (a_2 - b_2)^2}$$

In geography, other distance metrics are also used, such as the Manhattan metric:

$$d_M(a,b) = |a_1 - b_1| + |a_2 - b_2|,$$

which measures the distance traveled vertically and horizontally in a grid of blocks, as in a city with clearly defined blocks (i.e. as in Manhattan, New York).

As several geographic phenomena and processes are described by functions, metrics can also be induced on functions: e.g. for the distance between two continuous functions f,g: $[a,b] \in R$, the distance metric is:

$$d(f,g) = \int_a^b |f(x) - g(x)|dx$$

measuring the area between the graphs of f and g, from $x = a$ to $x = b$.

However, it has to be noticed that *not* all topological spaces can be induced by a metric.

Ultrametric spaces are governed by a peculiar ("ultrametric") topology, in which the usual Archimedean (triangle) inequality does not apply. Another stronger relationship applies instead, which is called "ultrametric":

$$|x+y|_p \leq \max\left(|x|_p, |y|_p\right) \leq |x|_p + |y|_p$$

So if three points x,y,z are the vertices of a triangle, the distances among them obey the following "non-Archimedean" relationship:

$$d(x,z) \leq \max(d(x,y), d(y,z)) \leq d(x,y) + d(y,z)$$

and hence, in such a space, all triangles are isosceles.

We live in the era of connectivity. Lines of coordinates invisibly associate to our lifelines, as well as lines of geographical coordinates or geodetic locations used for "presentity" identification. Analyses of social networks have documented the interplay between the geo-space and the cyber-space resulting from the exploitation of crowdsourced maps and platforms, such as OpenStreetMap (Neis et al., 2011; Corcoran & Mooney, 2013), Foursquare (Agryzkov et al., 2017; Chen et al., 2020) and Twitter (Agryzkov et al., 2016). Unsurprisingly perhaps, the overwhelming majority of the interactions in social networks is among friends (or even simply known to one another).

As suggested by Watts et al. (2002) in their influential paper, *friendships are ultrametric relationships*. The main reason justifying this assertion is that social distances are intransitive: the fact that two persons (i.e. A and B) may be friends does not imply that if A is also a friend with a third person (C) then B is also a friend of C. This intransitivity is characteristic of ultrametric spaces (Centola, 2015). So any two persons can be friends in some context but not in another and if they are, then they may not be so in any other contexts. And it is for this reason that ultrametric topology is essentially the topology of friendships and this is why it has been used to explain social networks (Schweinberger & Snijders, 2003), scientific networks (Stirling, 2007) and cultural interrelationships (Stivala et al., 2014).

Hierarchical structures as those described by ultrametric topologies abound in human and animal societies. All these connections are dendritic, with isosceles triangles and ultrametric topologies are perfectly suited to model landscape functions (Papadimitriou, 2013) so it is not a coincidence that ultrametric topology has found several applications in biogeography and biodiversity (Poulin et al., 2011; Oswald et al., 2016; Toni, 2021), remote sensing (Bradley et al., 2018), social networks analysis (Watts et al., 2002; Schweinberger & Snijders, 2003; Hua & Hovestadt, 2021; Costa et al., 2023) and modeling cultural dynamics (Nerbonne et al., 2008; Tronholm et al., 2012; Valori et al., 2012; Buechel et al., 2014; Stivala et al., 2014; Băbeanu et al., 2018).

The fact that all triangles are isosceles in ultrametric topologies has the counter-intuitive geometric implication that every point of a circular disk is at the center of the disk. The traditional topology of openness vs. closedness does not apply in ultrametric topology; the "clopen" topology applies to such disconnected entities instead (Fig. 10.1).

Similarly in 3D, two spheres are either one inside the other, or they do not intersect at all. This property is reminiscent of the notions of "*situatedness*" and "*centeredness*": the sense of place is closely related to the "situatedness of identity and action" and "the significance of place in modern life is associated with the fact, that as actors, we are always situated in place and period and that the contexts of our actions contribute to *our sense of centeredness*" (Entrikin, 1991, p.3–4). Thus

Fig. 10.1 A schematic representation of individualities in an ultrametric space (created by the author)

emerges a global cyber-physical environment of ubiquity as the world has become "*hypercentrique*" (Virilio, 2007, p.32, 72).

Yet, there is structure to this plurality of autonomous individuals and cyber-locations, because they both observe a hierarchical, dendroid structure, defined by their "ultrametric topology", in which the apexes meet at hierarchically higher levels. The Internet itself has an underlying hierarchical structure with its most important nodes (servers) atop of the hierarchy, followed by servers of medium and lower importance, down to personal blogs.

To illustrate an application of "ultrametric thinking" to the analysis of friendships among persons in the geo-space (the physical space) and in the cyber-space (i.e. through social networks, email exchanges etc), consider an example urban setting (Fig. 10.2) in which nine persons (coded from A to I) dwell in different buildings of the town. Some of them are friends in the cyber-space only (i.e. H and G), some only in the geo-space (i.e. E, G), while some of them are in both spaces (e.g. A and D).

Ultimately, all nodes are related in the total combined space, the GC-space (the combined geo-&-cyber-space; not to be confused with spaces of "mixed reality" or "extended reality"). In this fused cyber-physical setting, all triangles are isosceles, since either there is a friendship between any two persons or there is not (and this applies to both the geo-space and the cyber-space). Further, the relationships are represented by isosceles triangles because distances do not matter in either spaces (only the presence or absence of social connection or friendship between any two

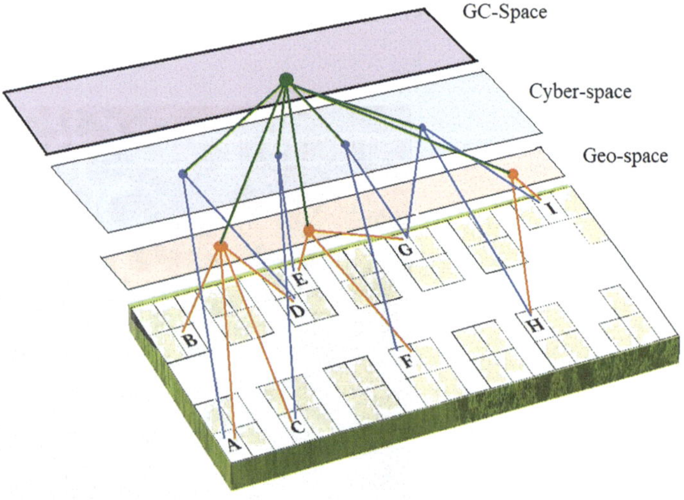

Fig. 10.2 An example urban setting, in which some individuals (coded by letters A to I with their locations marked at their home apartments) have developed friendships in the geo-space or in the cyber-space, or in both. All connections among nodes at these levels converge to a higher level, the combined GC-space

persons). Hence, the ultrametric distances between any two persons in the GC-space are

$$\begin{pmatrix}
 & A & B & C & D & E & F & G & H & I \\
A & - & 2 & 2 & 2 & 4 & 4 & 4 & 4 & 4 \\
B & & - & 2 & 2 & 2 & 4 & 4 & 4 & 4 \\
C & & & - & 2 & 2 & 4 & 4 & 4 & 4 \\
D & & & & - & 4 & 4 & 4 & 4 & 4 \\
E & & & & & - & 2 & 2 & 4 & 4 \\
F & & & & & & - & 2 & 4 & 4 \\
G & & & & & & & - & 2 & 2 \\
H & & & & & & & & - & 2 \\
I & & & & & & & & & -
\end{pmatrix}$$

Evidently, these are different than their distances in the geo-space:

$$\begin{pmatrix}
 & A & B & C & D & E & F & G & H & I \\
A & - & 2 & 2 & 2 & 4 & 4 & 4 & 4 & 4 \\
B & & - & 2 & 2 & 4 & 4 & 4 & 4 & 4 \\
C & & & - & 2 & 4 & 4 & 4 & 4 & 4 \\
D & & & & - & 4 & 4 & 4 & 4 & 4 \\
E & & & & & - & 2 & 2 & 4 & 4 \\
F & & & & & & - & 2 & 4 & 4 \\
G & & & & & & & - & 4 & 4 \\
H & & & & & & & & - & 2 \\
I & & & & & & & & & -
\end{pmatrix}$$

or in the cyber-space:

$$\begin{pmatrix}
 & A & B & C & D & E & F & G & H & I \\
A & - & 4 & 4 & 2 & 4 & 4 & 4 & 4 & 4 \\
B & & - & 4 & 4 & 4 & 4 & 4 & 4 & 4 \\
C & & & - & 4 & 2 & 4 & 4 & 4 & 4 \\
D & & & & - & 4 & 4 & 4 & 4 & 4 \\
E & & & & & - & 4 & 4 & 4 & 4 \\
F & & & & & & - & 2 & 4 & 4 \\
G & & & & & & & - & 2 & 2 \\
H & & & & & & & & - & 2 \\
I & & & & & & & & & -
\end{pmatrix}$$

The mean distances are: 132/36 = 3.67 for the cyber-space, 124/36 = 3.44 for the geo-space and 116/36 = 3.22 for the combined geo-cyber-space. Further, the Euclidean distance between the matrices of the geo-space and the GC is 4, the distance between those of the cyber-space and GC is 5.65 and that between cyber-space and geo-space is 6.32. Thus, in this fictitious geo-and-cyber-community, the higher proximity is in between the geo-space and the GC-space. From this example, it can also be seen that while the sum of ultrametric distances in the cyber-space is 132 and in the geo-space 124, in the GC-space it is less than both (116). This implies that, in this example, the total social distances may shrink if measured at a higher hierarchical level than those of the geo-space or the cyber-space. Or, the total social distances among people are shorter in the combined interactions in *both* the geo-space and the cyber-space levels (that may not apply to this particular example only).

References

Agryzkov, T., Martí, P., Nolasco-Cirugeda, A., Serrano-Estrada, L., Tortosa, L., & Vicent, J. F. (2016). Analysing successful public spaces in an urban street network using data from the social networks foursquare and twitter. *Applied Network Science, 1*, 1–15.

Agryzkov, T., Martí, P., Tortosa, L., & Vicent, J. F. (2017). Measuring urban activities using foursquare data and network analysis: A case study of Murcia (Spain). *International Journal of Geographical Information Science, 31*(1), 100–121.

Băbeanu, A. I., van de Vis, J., & Garlaschelli, D. (2018). Ultrametricity increases the predictability of cultural dynamics. *New Journal of Physics, 20*(10), 103026.

Bradley, P. E., Keller, S., & Weinmann, M. (2018). Unsupervised feature selection based on ultrametricity and sparse training data: A case study for the classification of high-dimensional hyperspectral data. *Remote Sensing, 10*(10), 1564.

Buechel, B., Hellmann, T., & Pichler, M. M. (2014). The dynamics of continuous cultural traits in social networks. *Journal of Economic Theory, 154*, 274–309.

Centola, D. (2015). The social origins of networks and diffusion. *American Journal of Sociology, 120*(5), 1295–1338.

Chen, Y., Hu, J., Xiao, Y., Li, X., & Hui, P. (2020). Understanding the user behavior of foursquare: A data-driven study on a global scale. *IEEE Transactions on Computational Social Systems, 7*(4), 1019–1032.

Corcoran, P., & Mooney, P. (2013). Characterising the metric and topological evolution of OpenStreetMap network representations. *The European Physical Journal Special Topics, 215*(1), 109–122.

Costa, F. X., Correia, R. B., & Rocha, L. M. (2023, January). The distance backbone of directed networks. In *Complex networks and their applications XI: Proceedings of the eleventh international conference on complex networks and their applications: Complex networks 2022—Volume 2* (pp. 135–147). Springer.

Cubitt, S. (1998). *Digital aesthetics*. Sage.

Entrikin, J. N. (1991). *The Betweenness of place*. Macmillan.

Hua, H., & Hovestadt, L. (2021). P-adic numbers encode complex networks. *Scientific Reports, 11*(1), 17.

Lee, Y., Jung, Y. J., Nam, K. W., Nittel, S., Beard, K., & Ryu, K. H. (2012). Geosensor data representation using layered slope grids. *Sensors, 12*(12), 17074–17093.

Neis, P., Zielstra, D., & Zipf, A. (2011). The street network evolution of crowdsourced maps: OpenStreetMap in Germany 2007–2011. *Future Internet, 4*(1), 1–21.

Nerbonne, J., Kleiweg, P., Heeringa, W., & Manni, F. (2008). Projecting dialect distances to geography: Bootstrap clustering vs. noisy clustering. In *Data analysis, machine learning and applications: Proceedings of the 31st annual conference of the Gesellschaft für Klassifikation eV, Albert-Ludwigs-Universität Freiburg, March 7–9, 2007* (pp. 647–654). Springer.

Nittel, S., Stefanidis, A., Cruz, I., Egenhofer, M., Goldin, D., Howard, A., et al. (2004). Report from the first workshop on geo sensor networks. *ACM SIGMOD Record, 33*(1), 141–144.

Oswald, J. A., Burleigh, J. G., Steadman, D. W., Robinson, S. K., & Kratter, A. W. (2016). Historical climatic variability and geographical barriers as drivers of community composition in a biodiversity hotspot. *Journal of Biogeography, 43*(1), 123–133.

Papadimitriou, F. (2013). Mathematical modelling of land use and landscape complexity with Ultrametric topology. *Journal of Land Use Science, 8*(2), 234–254.

Poulin, R., Krasnov, B. R., & Mouillot, D. (2011). Host specificity in phylogenetic and geographic space. *Trends in Parasitology, 27*(8), 355–361.

Reis, I. A. (2005). *Alternatives for Geosensor networks data analysis* (pp. 94–106). GeoInfo.

Schweinberger, M., & Snijders, T. A. (2003). Settings in social networks: A measurement model. *Sociological Methodology, 33*(1), 307–341.

Stirling, A. (2007). A general framework for analysing diversity in science, technology and society. *Journal of the Royal Society Interface, 4*(15), 707–719.

Stivala, A., Robins, G., Kashima, Y., & Kirley, M. (2014). Ultrametric distribution of culture vectors in an extended Axelrod model of cultural dissemination. *Scientific Reports, 4*(1), 1–9.

Toni, B. (2021). Archimedean and non-Archimedean approaches to mathematical modeling. *The mathematics of patterns, symmetries, and beauties in nature*. In *Honor of John Adam* (pp. 117–142). Springer Nature.

Tronholm, A., Leliaert, F., Sansón, M., Afonso-Carrillo, J., Tyberghein, L., Verbruggen, H., & De Clerck, O. (2012). Contrasting geographical distributions as a result of thermal tolerance and long-distance dispersal in two allegedly widespread tropical brown algae. *PLoS One, 7*(1), e30813.

Valori, L., Picciolo, F., Allansdottir, A., & Garlaschelli, D. (2012). Reconciling long-term cultural diversity and short-term collective social behavior. *Proceedings of the National Academy of Sciences of the USA, 109*, 1068–1073.

Virilio, P. (2007). *L'Université du désastre*. Galilée.

Watts, D. J., Dodds, P. S., & Newman, M. E. (2002). Identity and search in social networks. *Science, 296*(5571), 1302–1305.

Part V
Psychological, Educational, Epistemological and Philosophical Perspectives on Geo-Topology

Chapter 11
Geo-topology and Visual Impact

For the simplicity on this side of complexity,
I wouldn't give you a fig.
But for the simplicity on the other side of complexity,
for that I would give you anything I have.
(Oliver Wendell Holmes Jr., 1841–1935, "Holmes-Pollock Letters")

11.1 Contribution of Topology in Visual Impact Assessment

A significant part of the literature of psychology and cognitive sciences has focused on deciphering how shapes and forms are perceived by humans and the theory of Hochberg and McAlister (1953) for the understanding of 3D shapes is one of the prominent ones in this research field. Their formula for the visual impact V of a 3D shape focuses on three fundamental topological and geometric characteristics of shapes: line segments (L), junction points (J) and angles (A) and connects all these by a simple sum: $V = L + J + A$. Hochberg & McAlister (1953) defined the visual impact of shapes as perceived by humans and their formula has been used in numerous psychological experiments ever since, also extending to polyhedra (Hochberg & Brooks, 1960; Hochberg, 1978). With further elaborations of the Hochberg and McAlister theory by cognitive psychologists (Perkins, 1972, 1976; Watt, 1984; Poggio et al., 1985; Lowe, 1972; Rock & DiVita, 1987; Hekkert & van Wieringen, 1990; Gordon, 1997; Regan, 2000; Todd & Norman, 2003) and a significant wealth of applications has been accumulated that turned out to be useful even in various research fields, i.e. computer vision (Boyer & Sarkar, 1999, 2000).

The visual impact of landscapes is usually examined experimentally on the basis of questionnaires or by considering statistical analyses of values of indices of landscape structure such as landscape diversity, or both (Patsfall et al., 1984; Ewald, 2001; Daniel & Meitner, 2001; Hagerhall et al., 2004; Kühne, 2006, 2012, 2018, 2019b; Benson, 2008; Singh et al., 2008; Kühne & Weber, 2019). Also, it can be examined by adopting phenomenological approaches to landscape analysis with emphasis on individual modes of perception (Kühne, 2019a) in contrast to positivist approaches (Weber & Kühne, 2019). This is because landscapes are not simply

© The Author(s), under exclusive license to Springer Nature Switzerland AG 2023
F. Papadimitriou, *Geo-Topology*, GeoJournal Library 133,
https://doi.org/10.1007/978-3-031-48185-7_11

physical entities; in fact, there are invisible flows of energy and information eventually reflecting a "threefold landscape dynamics" that extends well beyond the realm of nature (Kühne & Jenal, 2020). Further, aesthetic experience might also relate to the development of ethical attitudes towards the environment (Brady, 2006). Yet, methods and concepts of topology have already been connected to aesthetics (Parisi, 2012; Cazeaux, 2013; Girot, 2013; Lara-Cabrera et al., 2014; Dapogny et al., 2017; Papadimitriou, 2020b; Carballal et al., 2021; Loos et al., 2022; Papadimitriou, 2022d).

Translating the Hochberg and McAlister formula to 2D entities such as digital images, each one of its terms can be defined topologically by making use of the topological boundary operator $\partial(.)$ of (n-1)-dimensional boundaries of n-dimensional simplexes (see Papadimitriou, 2020a, b). Either in its (original) 3D form or in the 2D form (proposed here), the Hochberg and McAlister theory is based on a combined topological and geometric approach, since the boundary of a 2D area is its line in case of a grid and the boundary of a 1D line are its two (zero-dimensional) junction points, which, in the case of the 2D model, are simply the intersection points of the grid lines (Fig. 11.1).

Consequently, the model defined at the cell level both topologically (in two of its parameters) and geometrically (the third parameter) depends on boundaries and angles only:

- The sum of lines (L), resulting from the action of the topological boundary operator $\partial_i(.)$ on the simplex (square) R with respect to its four surrounding square cells. The boundary between two 2D adjacent regions defined on a planar grid ($n = 2$) is an n-1 entity, that is the 1D entity "line".
- The sum of junction points (J): although not all cells have junction points to be considered in the calculation of the visual impact, the junction points nevertheless

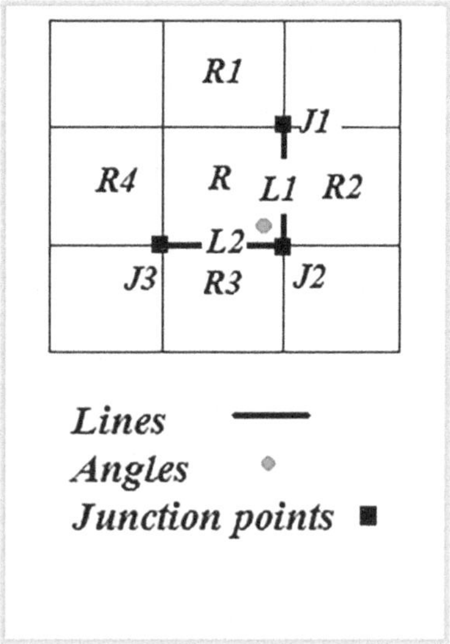

Fig. 11.1 Definition of the three basic variables of visual impact J,A,L for a region R of a raster map

result from the application of the boundary operator $\partial_j(.)$ on each one of the lines (L) previously defined, since the boundary of an 1D segment is the set of the two points defining it (0D entities).
– The sum of angles (A) defined by the lines (L) and included in each cell (this is a geometric variable).

In order to standardize the values of V, the maximum value of V needs to be calculated (V_{max}), whereas the minimum V is simply zero (corresponding to $L = J = A = 0$). The maximum value of V can be calculated from the maximum values of L, J, A, which for any map with r rows and c columns are:

$$L = c(r+1) + r(c+1)$$
$$J = (r+1)(c+1)$$
$$A = 4rc$$

and hence, the maximum V is:

$$V_{max} = (r+1)(c+1) + 4rc + r(c+1) + c(r+1)$$
$$= (r+1)(2c+1) + 5rc + 1$$

In case of square maps ($r = c$), these variables simplify to

$$L = 2c(c+1)$$
$$J = (c+1)^2$$
$$A = 4c^2$$

and thus

$$V_{max} = 7c^2 + 4c + 1.$$

Consequently, for a square map, the measure of the normalized visual impact is:

$$V_{norm} = \frac{V}{V_{max}} = \frac{V}{7c^2 + 4c + 1}$$

The area of Leblon (Southern Rio de Janeiro) was selected for the application of this formula; a region with forests, built up lands, including the lake Rodrigo de Freitas, and other land uses (athletic installations). The landscape of Rio de Janeiro has high diversity, large protected areas (which are mostly montane Atlantic rainforests) and its management is a highly complex process. The impact of formal and informal urbanization (Macedo, 2007) has significant impact on the visual amenities of this landscape, while almost half of the protected areas overlap with areas covered by the urban sprawl (Uehara-Prado & Fonseca, 2007; Chazdon et al., 2009). As a result, less than 90% of the extent of the original Brazilian Atlantic forest has been left intact nowadays, while some areas are still being logged and this has significant adverse ecological impacts on the area (Harris et al., 2005; Villela et al., 2006). The sensitive relationships between peri-urban forests and urbanization have prompted for the application of various conservation initiatives (Pedlowski et al., 2002; Scarano, 2002). Landscape aesthetics is one of the criteria for landscape management, due to the extraordinary beauty of these landscapes (Papadimitriou, 2012). The satellite image was first reclassified using GIS (Fig. 11.2). Larger quadrants are created by re-grouping the cells of the initial 8x8 map so as to derive a 4x4 map so the values of *J,A,L* can be calculated thus yielding the values of *V* for the 16 quadrants (Fig. 11.3).

Having defined the boundaries among the different land use types, the junctions *J* were marked at inflection points and the calculation of the number of angles *A* followed next. No boundary lines were measured on the outer border of the landscape map, while angles were counted only if they touched any of the quadrants' junction points (Fig. 11.3).

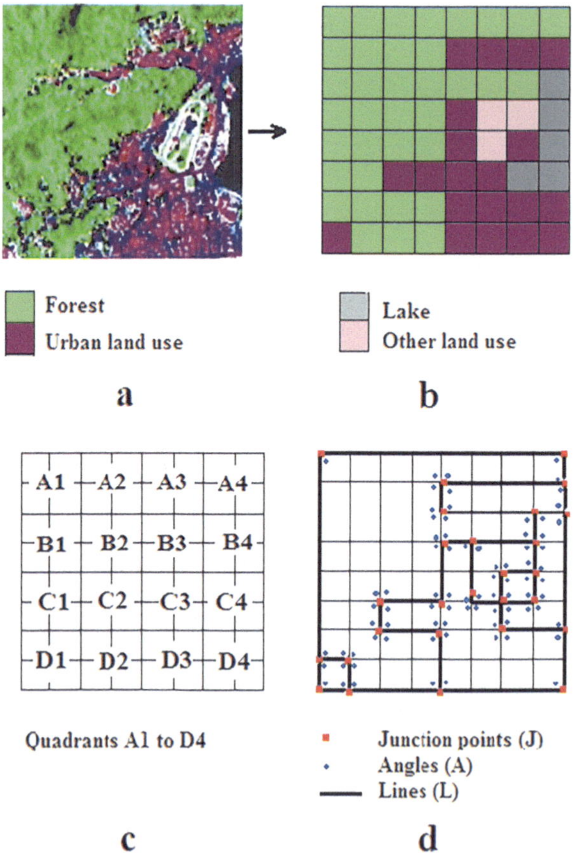

Fig. 11.2 The landscape of Leblon area, Rio de Janeiro from satellite image (**a**) and its reclassification in 8×8 raster format using GIS (**b**). Quadrants of the rasterized image that resulted from the re-grouping the cells of the 8×8 map (**c**) and the identification of the topological-geometric entities for the calculation of the visual impact (**d**)

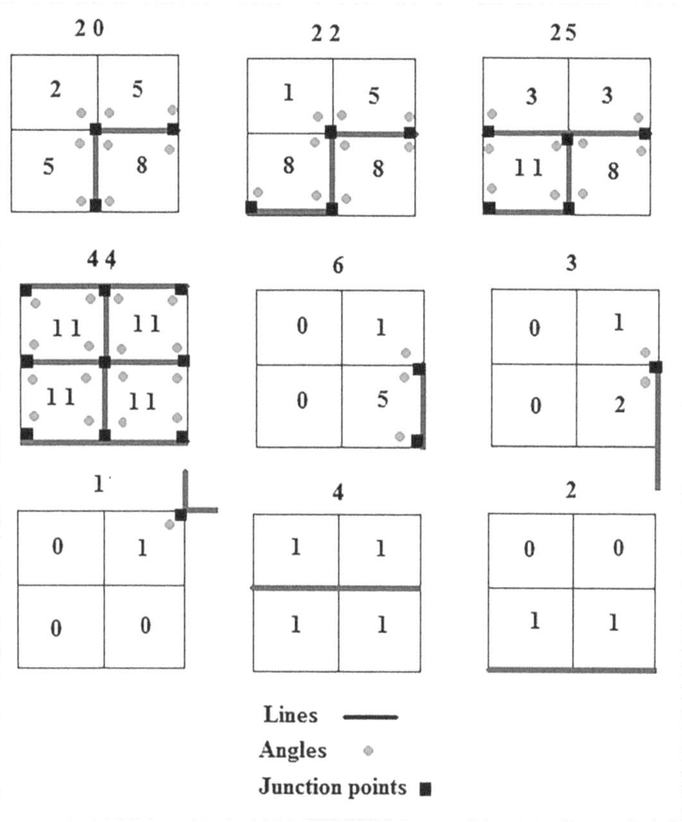

Fig. 11.3 Example calculations of the visual impact V on various possible configurations of quadrants, with lines, junction points and angles on raster formats. The values of V per cell are shown along with the values of V of each quadrant

Following the results of the calculations of V per 2x2 quadrant (table 11.1), the mean V is 11.81 (with s.d. = 6.83). However, considering that the maximum visual impact is $V_{max} = (7 \times 8^2) + (4 \times 8) + 1 = 481$, the average visual impact (for all 16 quadrants) is:

$$V_{norm(landscape)} = \frac{\sum_{Q=1}^{Q=16} V_Q}{V_{max}} = \frac{189}{481} = 39.29\%$$

of the maximum possible.

By correlating the visual impact with land cover/land use types, it is revealed that the land use types “other” (the sportsgrounds) and “lake” (Rodrigo de Freitas) correspond to highest values of visual impact (Fig. 11.4).

Table 11.1 Calculation of visual impact per 2x2 quadrant

Quadrants	Lines (L)	Angles (A)	Junctions (J)	Visual impact $V = L + A + J$
A_1	2	1	1	4
A_2	2	3	2	7
A_3	4	3	2	9
A_4	4	4	4	12
B_1	1	0	0	1
B_2	1	3	2	6
B_3	5	6	3	14
B_4	7	10	5	22
C_1	2	3	2	7
C_2	4	6	4	14
C_3	5	11	6	22
C_4	7	11	6	24
D_1	6	10	5	21
D_2	3	3	3	9
D_3	2	3	3	8
D_4	3	3	3	9
Total				189

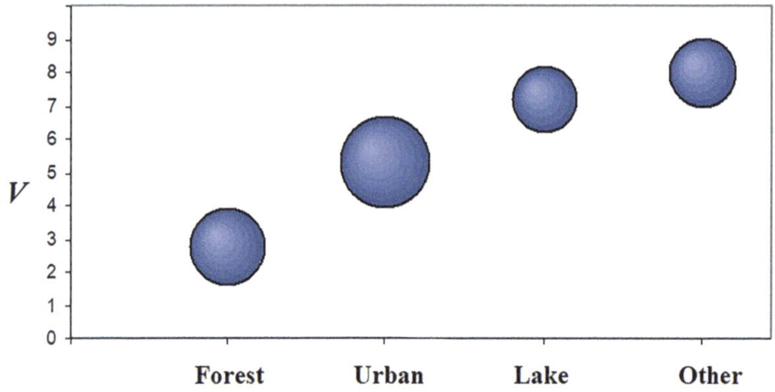

Fig. 11.4 Average visual impact per land cover/land use (ball radii are equal to the standard deviation) for all 64 cells of the image. The landscape types "lake" and "other land uses" relate with higher visual impact than "urban" and "forest"

11.2 Topological Boundaries, Visual Impact and Entropy

After assessing the visual impact on the basis of the Hochberg and McAlister formula, it is possible to correlate it with other also topologically derived indices of landscape structure. One such is the number of boundaries between cells on a raster map with c = columns and r = rows and u = total number of land uses present on the landscape ($u = 1,2,3, \ldots ,m$):

$$k = 2cr - (c + r),$$

Cells are 2D entities and their boundaries are 1D (lines), so the sum of boundaries of each land use type A_i with itself (internal line boundaries on a grid) is defined as

$$w_{ii} = \sum_{i=1}^{u} \partial A_{i,i}$$

while the sum of boundaries of each land use type A_i with other land use types (A_j) is

$$w_{ij} = \sum_{i=1}^{u} \sum_{\substack{j=1 \\ i \neq j}}^{u} \partial A_{i,j}.$$

Consequently, a Shannon-type metric of entropy of boundaries H_B can be defined as (Papadimitriou, 2022b):

$$H_B = \sum_{i=1}^{u} \left(\frac{w_{ii}}{k}\right) \log\left(\frac{k}{w_{ii}}\right) + \sum_{i=1}^{u} \sum_{\substack{j=1 \\ i \neq j}}^{u} \left(\frac{w_{ij}}{k}\right) \log\left(\frac{k}{w_{ij}}\right).$$

with $max(H_B) = \log(m)$, where m = maximum number of land uses that the landscape could possible host (occuring only when each cell is occupied by a different land use type) and $min(H_B) = 0$, which is realized when the entire landscape has only one land use ($u = 1$) and this condition yields $\sum_{i=1}^{u} \partial A_{i,i} = 0$.

As it turns out, all correlations of visual impact V with the values of spatial entropies H_1, H_3, H_4, H_B (Papadimitriou, 2022a, b) for the 16 quadrants of size 2 × 2 of the image are statistically significant at $p < .01$ (Table 11.2) and hence the visual impact of this landscape is significantly correlated with its spatial entropy.

This measure can be applied to the landscape map of Leblon (Fig. 11.5), from which it follows that V agrees with the entropy of boundaries and with the spatial entropy measure H_3 in all 16 quadrants except for one (the quadrant A_3).

Table 11.2 Results of the statistical analysis of the relationships between visual impact (V) and spatial entropies H_1, H_3 and H_4

Spatial entropy	Pearson's R	P-value	Significant at
H_1	0.7137	.001904	$p < .01$
H_3	0.7251	.001448	$p < .01$
H_4	0.7261	.001448	$p < .01$
H_B	0.6981	.002635	$p < .01$

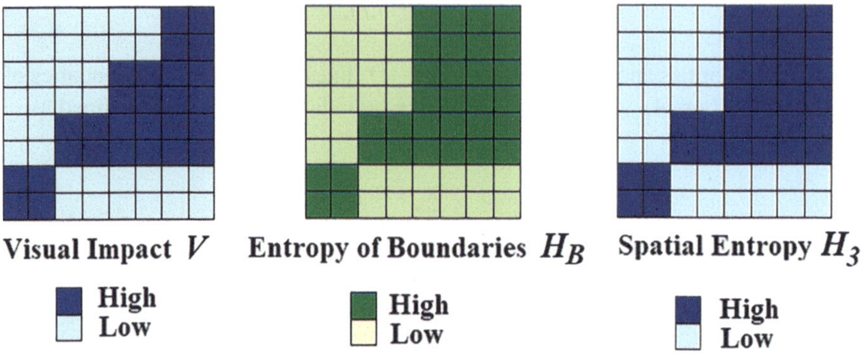

Fig. 11.5 The values of high and low visual impact V coincide with those of the entropy of boundaries H_B and with the spatial entropy measure H_3

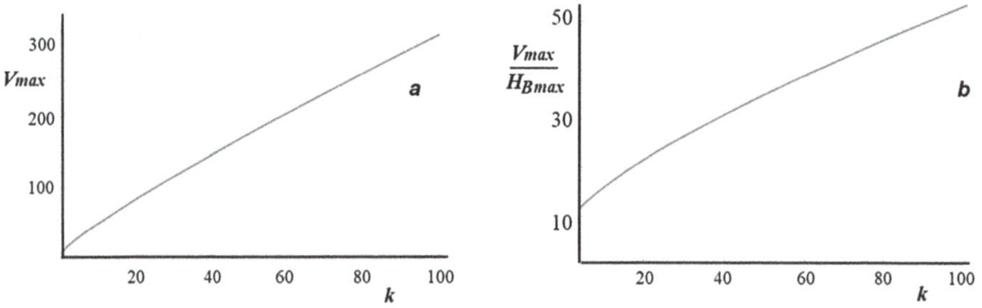

Fig. 11.6 Change of the visual impact V and change with respect to the maximum entropy of boundaries H_B against k (the number of boundaries between cells)

After some algebraic manipulations, the maximum visual impact V_{max} and the entropy of boundaries H_B can be derived only from the total number of boundaries k:

$$V_{\max} = \frac{7k + 11\sqrt{2k+1} + 6}{2}$$

and

$$H_B = \log_2\left(\frac{k + \sqrt{2k+1} + 4}{2}\right)$$

from which it follows that the correlation between the entropy of boundaries H_{Bmax} and k appears to be almost linear, as does the ratio $Vmax/H_{Bmax}$ (Fig. 11.6).

Also, the directional entropy (Papadimitriou 2022c) of the visual impact V for the entire image is $H_{\text{dir}} = 3.5212$ (Fig. 11.7) that is $H_{\text{dir}}/H_{\text{dirmax}} = 85\%$ of the maximum H_{dir} (which results from the uniform distribution around the circle). As can be

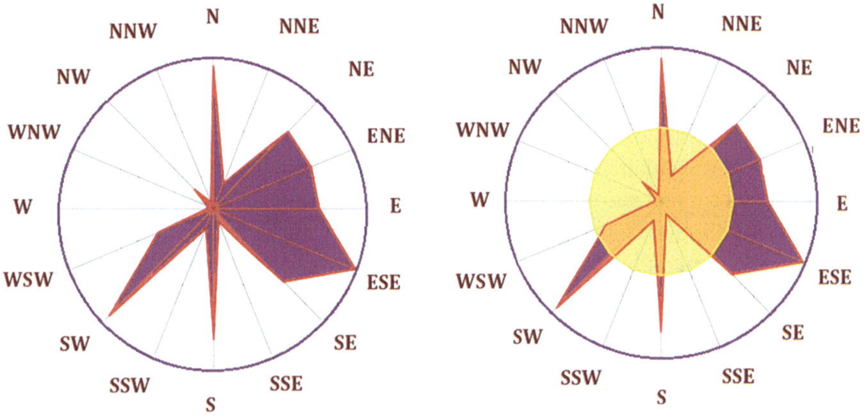

Fig. 11.7 The H_{dir} (left) calculated from angular sectors of the Leblon image and compared to the maximum H_{dir} that would be expected from a uniform distribution (right) reveals that the eastern side of the image has higher spatial entropy

verified, the eastern half of the landscape has a higher spatial entropy (either if measured as H_3 or H_{dir}), therefore corroborating the assessment on the basis of V that the eastern areas of the landscape have higher visual impact.

An alternative formulation of the model might be derived without counting the external borderline of the landscape. In this case, the equations for the borderless square model become:

$$L = 2c(c - 1)$$
$$J = (c + 1)^2 - 4$$
$$A = 4c^2 - 4$$

and hence,

$$V_{w\,\max} = 7\left(c^2 - 1\right)$$

in which case, the visual impact is lower (by $4c$-6) than that of the bordered model.

The adjustment of the Hochberg and McAlister formula to measure the visual impact of landscapes (or maps of landscapes) is an example of how some very basic notions of topology can be used to tackle a problem which has hitherto been approached only statistically (i.e. by analyzing replies to questionnaires and combinations of structural landscape indices). Evidently, non-topological approaches are not expected to have any general applicability due to their being inherently local, regional, or site-dependent. Interestingly, while research in experimental psychology has suggested that many people would prefer images of lower visual impact (Van der Helm & Leeuwenberg, 1996; Van der Helm et al., 1992), there is evidence that other people tend to prefer complex rather than simple structures (Reber et al., 2004).

We probably need to gain a deeper understanding of the relationships among topology, visual impact and complexity of geographical imagery. In later experiments using objects of varying complexity for instance (Norman et al., 2010), it was shown that either very simple or very complex objects were more preferred. Whatever the relationship of the measure *V* with aesthetics and spatial complexity might be and whether evaluations of *V* are more suitable for the evaluation of the potential of landscapes for recreation or for assessing conservation needs, topological models based on the simplest topological constituents (with the exception of angles in this case) of the geographical space and its representations (satellite imagers, maps etc) may be considered as objective indicators of the visual impact of landscapes.

References

Benson, J. (2008). Aesthetic and other values in the rural landscape. *Environmental Values, 17*(2), 221–238.

Boyer, K. L., & Sarkar, S. (1999). Perceptual organization in computer vision: Status, challenges and potential. *Computer Vision & Image Understanding, 76*, 1.

Boyer, K. L., & Sarkar, S. (2000). *Perceptual organization for artificial vision systems*. Kluwer.

Brady, E. (2006). Aesthetics in practice: Valuing the natural world. *Environmental Values, 15*(3), 277–291.

Carballal, A., Cedron, F., Santos, I., Santos, A., & Romero, J. (2021). Minimal neural network topology optimization for aesthetic classification. *Neural Computing and Applications, 33*, 107–119.

Cazeaux, C. (2013). Interrupting the artist: Theory, practice and topology in Sartre's aesthetics. In *Thinking through art* (pp. 40–50). Routledge.

Chazdon, R. L., Harvey, C., Komar, O., Griffith, D., Ferguson, B. G., Martinez-Ramos, M., Morales, H., Nigh, R., Soto-Pinto, L., Van Breugel, M., & Philpott, S. (2009). Beyond reserves: A research agenda for conserving biodiversity in human-modified tropical landscapes. *Biotropica, 41*(2), 142–153.

Daniel, T., & Meitner, M. M. (2001). Representational validity of landscape visualizations: The effects of graphical realism on perceived scenic beauty of forest vistas. *Journal of Environmental Psychology, 21*(1), 61–72.

Dapogny, C., Faure, A., Michailidis, G., Allaire, G., Couvelas, A., & Estevez, R. (2017). Geometric constraints for shape and topology optimization in architectural design. *Computational Mechanics, 59*, 933–965.

Ewald, K. C. (2001). The neglect of aesthetics in landscape planning in Switzerland. *Landscape and Urban Planning, 54*(1), 255–266.

Girot, C. (2013). The elegance of topology. *Landscript, 3*, 79–116.

Gordon, I. E. (1997). *Theories of visual perception*. Wiley.

Hagerhall, C. M., Purcell, T., & Taylor, R. (2004). Fractal dimension of landscape silhouette outlines as a predator of landscape preference. *Journal of Environmental Psychology, 24*(2), 247–255.

Harris, G. M., Jenkins, C. N., & Stuart, P. (2005). Refining biodiversity conservation priorities. *Conservation Biology, 19*(6), 1957–1968.

Hekkert, P., & van Wieringen, P. C. W. (1990). Complexity and prototypicality as determinants of appraisal of cubist paintings. *British Journal of Psychology, 81*, 483–495.

Hochberg, J. (1978). *Perception*. Prentice Hall.

Hochberg, J., & Brooks, V. (1960). The psychophysics of form. Reversible-perspective drawings of spatial objects. *American Journal of Psychology, 73*, 337–354.

Hochberg, J. E., & McAlister, E. (1953). A quantitative approach to figural "goodness". *Journal of Experimental Psychology, 46*, 361–364.
Kühne, O. (2006). *Landschaft in der Postmoderne: das Beispiel des Saarlandes*. DUV.
Kühne, O. (2012). *Stadt-Landschaft-Hybridität: Ästhetische Bezüge im postmodernen Los Angeles mit seinen modernen Persistenzen*. Springer VS.
Kühne, O. (2018). *Landscape and power in geographical space as a social-aesthetic construct*. Springer.
Kühne, O. (2019a). Phänomenologische Landschaftsforschung. In O. Kühne, F. Weber, K. Berr, & C. Jenal (Eds.), *Handbuch Landschaft* (pp. 135–144). Springer VS.
Kühne, O. (2019b). *Landscape theories: A brief introduction*. Springer VS.
Kühne, O., & Jenal, C. (2020). The threefold landscape dynamics: Basic considerations, conflicts, and potentials of virtual landscape research. In D. Edler, C. Jenal, & O. Kühne (Eds.), *Modern approaches to the visualization of landscapes* (pp. 389–402). Springer VS.
Kühne, O., & Weber, F. (2019). *Hybrid California: Annäherungen an den Golden State, seine Entwicklungen, Ästhetisierungen und Inszenierungen*. Springer VS.
Lara-Cabrera, R., Cotta, C., & Fernández-Leiva, A. J. (2014). Geometrical vs topological measures for the evolution of aesthetic maps in a RTS game. *Entertainment Computing, 5*(4), 251–258.
Loos, S., Wolk, S. V. D., Graaf, N. D., Hekkert, P., & Wu, J. (2022). Towards intentional aesthetics within topology optimization by applying the principle of unity-in-variety. *Structural and Multidisciplinary Optimization, 65*(7), 185.
Lowe, D. G. (1972). *Perceptual organization and visual recognition*. Kluwer.
Macedo, J. (2007). Urban land policy and new land tenure paradigms: Legitimacy vs. legality in Brazilian cities. *Land Use Policy, 25*(2), 259–270.
Norman, J. F., Beers, A. M., & Phillips, F. (2010). Fechner's aesthetics revisited. *Seeing and Perceiving, 23*(3), 263–271.
Papadimitriou, F. (2012). Modelling landscape complexity for land use Management in Rio de Janeiro, Brazil. *Land Use Policy, 29*(4), 855–861.
Papadimitriou, F. (2020a). *Spatial entropy and landscape analysis*. Springer.
Papadimitriou, F. (2020b). Exploring spatial complexity in 3d. In *Spatial complexity: Theory, mathematical methods and applications* (pp. 101–113). Springer.
Papadimitriou, F. (2022a). *Spatial complexity: Theory, mathematical methods and applications*. Springer.
Papadimitriou, F. (2022b). Computing the spatial entropy of square binary maps. In *Spatial entropy and landscape analysis* (pp. 15–30). Springer.
Papadimitriou, F. (2022c). Spatial entropy of directional geographical data and landscape networks. In *Spatial entropy and landscape analysis* (pp. 31–55). Springer.
Papadimitriou, F. (2022d). Visual perception of spatial entropy and landscape analysis. In *Spatial entropy and landscape analysis* (pp. 87–102). Springer.
Parisi, L. (2012). Digital design and topological control. *Theory, Culture & Society, 29*(4–5), 165–192.
Patsfall, M. R., Feimer, N. R., Buyhoff, G. J., & Wellman, J. D. (1984). The prediction of scenic beauty from landscape content and composition. *Journal of Environmental Psychology, 4*(1), 7–26.
Pedlowski, M. A., Da Silva, V. A. C., Adell, J. J. C., & Heynen, N. C. (2002). Urban forest and Environmental inequality in Campos dos Goytacazes, Rio de Janeiro, Brazil. *Urban Ecosystems, 6*(1–2), 9–20.
Perkins, D. N. (1972). Visual discrimination between rectangular and nonrectangular parallelopipeds. *Perception & Psychophysics, 12*, 396–400.
Perkins, D. N. (1976). How good a bet is good form? *Perception, 5*, 393–406.
Poggio, T., Torre, V., & Koch, C. (1985). Computational vision and regularization. *Nature, 317*, 314–319.

Reber, R., Schwarz, N., & Winkielman, P. (2004). Processing fluency and aesthetic pleasure: Is beauty in the Perceiver's processing experience? *Personality and Social Psychology Review, 8*(4), 364–382.

Regan, D. (2000). *Human perception of objects*. Sinauer.

Rock, I., & DiVita, J. (1987). A case of viewer-centered object perception. *Cognitive Psychology, 19*, 280–293.

Scarano, F. R. (2002). Structure, function and floristic relationships of plant communities in stressful habitsts marginal to the Brazilian Atlantic rainforest. *Annals of Botany, 90*(4), 517–524.

Singh, S. N., Donavan, T., & Mishra, S. (2008). The latent structure of landscape perception: A mean and covariance structure modelling approach. *Journal of Environmental Psychology, 28*(4), 339–352.

Todd, J. T., & Norman, J. F. (2003). The visual perception of 3D shape from multiple cues: Are observers capable of perceiving metric structure? *Perception & Psychophysics, 65*, 31–47.

Uehara-Prado, M., & Fonseca, R. L. (2007). Urbanization and mismatch with protected areas place the conservation of a threatened species at risk. *Biotropica, 39*(2), 264–268.

Van der Helm, P., & Leeuwenberg, E. (1996). Goodness of visual regularities: A non-transformational approach. *Psychological Review, 103*, 429–456.

Van der Helm, P., Van Lier, R., & Leeuwenberg, E. (1992). Serial pattern complexity: Irregularity and hierarchy. *Perception, 21*, 517–544.

Villela, D. M., Nascimento, M. T., Aragao, L. W., & da Gama, D. (2006). Effect of selective logging on forest structure and nutrient cycling in a seasonally dry Brazilian Atlantic forest. *Journal of Biogeography, 33*(3), 506–516.

Watt, R. J. (1984). Towards a general theory of the visual acuities for shape and spatial arrangement. *Vision Research, 24*, 1377–1386.

Weber, F., & Kühne, O. (2019). Essentialistische Landschafts-und positivistische Raumforschung. In *Handbuch Landschaft* (pp. 57–68). Springer VS.

Chapter 12
Geo-topology in Games and Education

As in a game of cards,
so in the game of life, we must play what is dealt to us,
and the glory consists, not so much in winning,
as in playing a poor hand well.
(Josh Billings, 1818–1885, "On Ice and other things")

12.1 Map Coloring Games

Using games in geographical education has manifold advantages in terms of learning new knowledge, fostering environmental values and sharpening spatial problem-solving skills (Walford, 1981; Cartwright, 2006; Da Silva, 2015; Kim & Shin, 2016; Fontaine, 2020; Papadimitriou, 2020, 2022; Morawski & Wolff-Seidel, 2023) and there are different ways by which concepts of topology can be explored within the context of geographical education and geo-educational games in particular.

Beginning with mapping exercises, a starting point may be the "Four Colors Theorem" which puzzled topologists for many years, but it can turn out to be a interesting for geographical education if converted into a mapping game.

The "chromatic number" of a map M is calculated from the number of its vertices V, edges E and regions F:

$$\chi(M) = V - E + F$$

Take, for instance, a simple vector map M_1 (Fig. 12.1a).

Applying the formula yields:

$$\chi(M_1) = 10 - 15 + 7 - 2$$

Certain types of connections may change the value of the chromatic number. If, for instance, one point is added in one of the regions of the map along with a self-connecting loop around it, then the resulting map M_2 (Fig. 12.1b) has the chromatic number

© The Author(s), under exclusive license to Springer Nature Switzerland AG 2023

F. Papadimitriou, *Geo-Topology*, GeoJournal Library 133,
https://doi.org/10.1007/978-3-031-48185-7_12

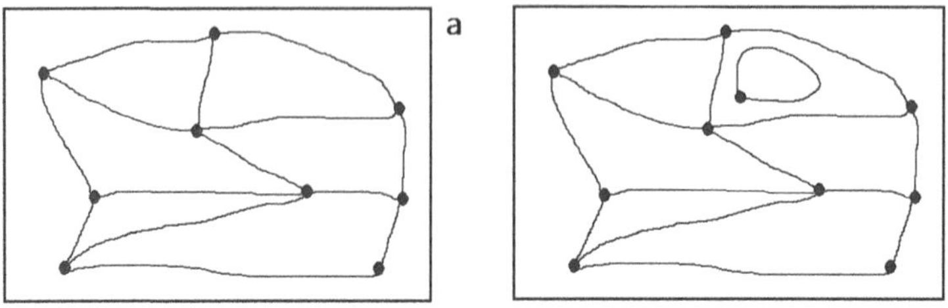

Fig. 12.1 A simple vector map M_1 (**a**) and the map M_2 in which a point and a loop have been added (**b**)

$$\chi(M_2) = 11 - 16 + 8 = 3$$

The chromatic number of a map varies according to the topology of the surface that the map is drawn on i.e. for maps on a cylinder it is 4, but it is 6 for the Möbius strip and for the Klein bottle.

Variations of mapping games using the Four Colors Theorem can be based on "Heffter's problem", which is to decide how to color maps in which each face is adjacent to every other one. Then, the problem is to calculate the least number g such that n neighboring regions may be constructed on a g-holed doughnut, in which case g is evaluated from the formula (Biggs et al., 1976):

$$g(n) > \left\lfloor \frac{(n-3)(n-4)}{12} \right\rfloor$$

On the surface of a sphere, the number of countries minus the number of boundary lines, plus the number of meeting points, is always 2: F-E + V = 2.

But, on the surface of a torus the result is zero and so every map drawn on a torus has at least one country with 6 or fewer neighboring countries. It thus follows that the "Four Colors Theorem" for maps on the surface of a torus should be adjusted to 7 instead of 4. Equivalently, every map on the surface of a 2-holed torus can be colored with 8 colors and, eventually, a map on the surface with h holes can be colored with 2-2 h colors. However, a more general map color theorem for a surface S without a boundary (i.e. without lakes) is given by Heawood's estimate of the "chromatic number" of the surface h(S) from its Euler number $\chi(S)$:

$$h(S) = \left\lfloor \frac{7 + \sqrt{49 - 24\chi(S)}}{2} \right\rfloor$$

which implies that such a surface will require no more than $h(S)$ colors. A map on a surface with 10 holes for instance, will require 14 colors. Yet, there are the so-called "empire maps" of countries with "colonies" outside their main territory. Heawood showed that 12 colors will suffice to color such maps (although there are exceptions, i.e. a 6-colored Klein bottle does not obey to Heawood's estimate).

Map coloring games can be created so that the selection of colors by which the regions are colored is important to color a map properly. For instance, students may verify that if a map is colored with the wrong sequence of steps, then they may (falsely) conclude that four colors are not enough (Fig. 12.2). Further to the Four Colors Theorem, games of mazes can be devised, prompting students to apply Jordan's curve theorem in order to decide whether a point in the maze is interior or exterior to the maze (Fig. 12.3).

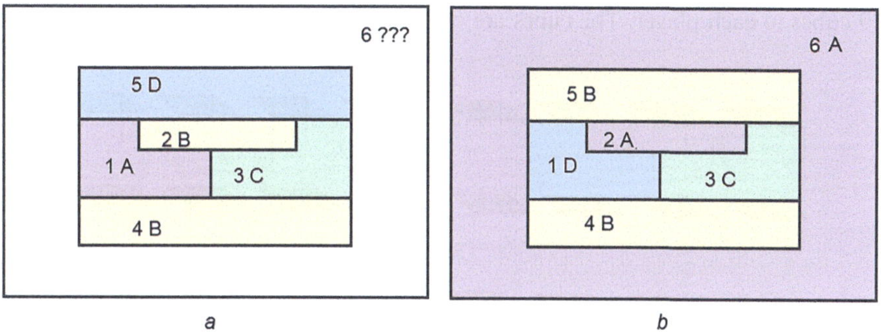

Fig. 12.2 A coloring game may be designed with the aim of determining the correct order of coloring in order to eventually color a map on the plane with no more than four colors. In this example, a wrong selection of colors (A,B,C,D) and steps (1,2,3,4,5,6) results in the sixth step requiring a fifth color (**a**). Yet, the correct selection of colors and steps (**b**) proves that the map can indeed be colored with four colors only

Fig. 12.3 Mazes can turn into exciting topological games and Jordan's curve theorem can be interesting to experiment with, aiming at deciding whether a point is internal or external to the maze; this can be quite a challenge if the maze is overly complex such as this one

12.2 Adjacency Games

Games may also be used in geographical and environmental education to advance knowledge in landscape management and conservation (Sandbrook & Monteferri, 2015; Swetnam & Korenko, 2019; Santos et al., 2020). So aside of map coloring games, the concept of topological adjacencies might be exploited for training in forming spatial arrangements that obey certain ecological/environmental rules. One such game is "Geokivi" (from the greek word "ΓεωΚύβοι", meaning earth-cubes); a board game created by the author, combining luck and strategy and intended for two players.

Geokivi uses 58 small cubes (Fig. 12.4) randomly divided in two by allocating 29 cubes to each player. The cubes are of two types: those with only one of the six

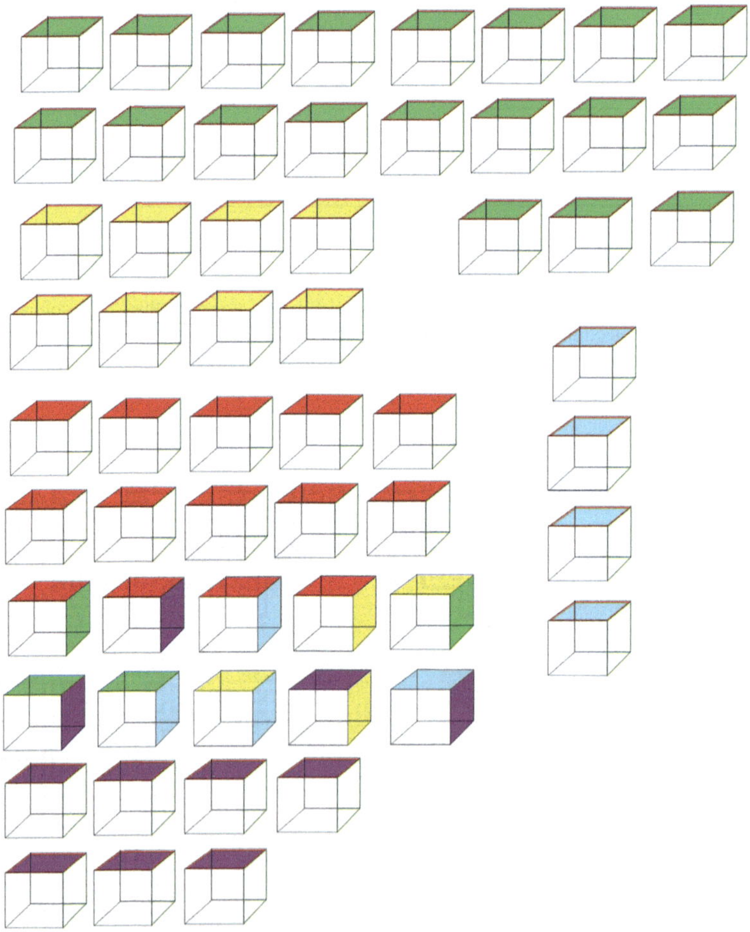

Fig. 12.4 The 58 cubes of Geokivi

faces painted and those with two faced painted. There are only five possible colors, with each color corresponding to a landscape type: red for urban/residential, green for forest, yellow for agriculture, indigo for industrial, and light blue for water bodies. Of the 58 cubes, 48 have only one face painted and 10 have two faces. Those with two faces painted have different colors on each one of their two painted faces (the other four faces remain unpainted), thus offering players the possibility to chose which one of the two colors (faces) to play every time.

Each player's goal is to create sets of four cubes in quartets, either on a row or in 2x2 squares, with these quartets consisting of at least three cubes of the same color and the fourth cube of compatible color. A player wins a point upon laying the last (fourth) cube that completes such a quartet. Two points are won when a player forms a cross-shaped quintet (either in 3x3 arrangements or on a row), with at least four cubes of the same color (Fig. 12.5).

The initial allocation of 29 cubes to each player is made by chance, but the players' strategy consists in deciding which cube to lay on the board every time and

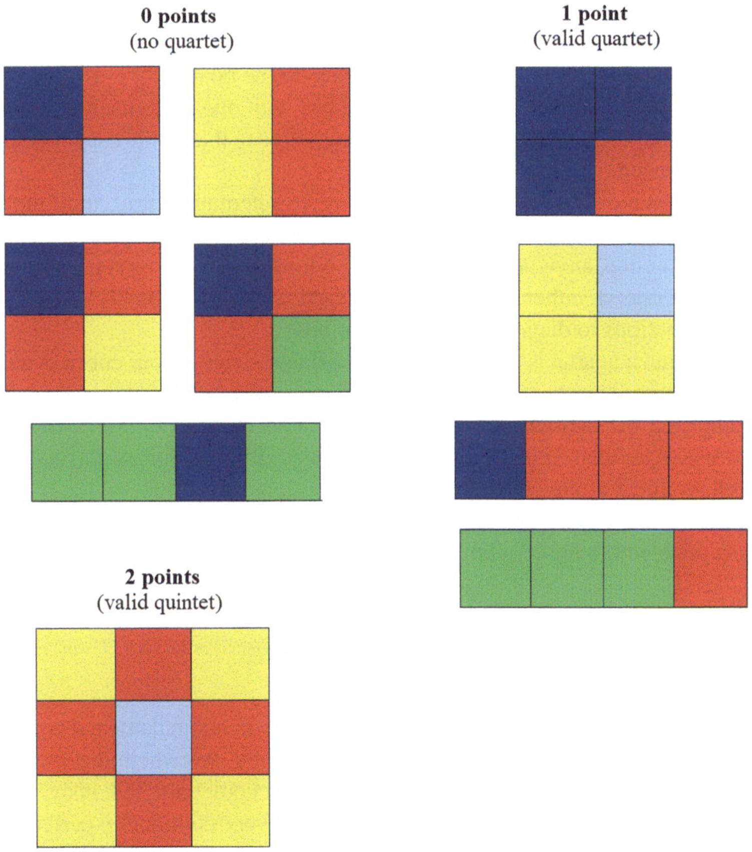

Fig. 12.5 Examples of permissible and non-permissible combinations in quartets, in a quintet and in rows

where, so as to win more points. The educational context consists in observing color compatibilities, so that each cube color is adjacent only with compatible cube colors. This compatibility simulates the creation of environmentally acceptable landscape settings, by following some permissible chromatic combinations only (Fig. 12.5), which reflect basic environmental protection precautions and are as follows:

- Urban land use (red) can be combined with any other land use.
- Forests (green) can be combined with all other land uses apart from industry, with which it can not be combined in any way (neither face-to-face nor diagonally).
- Agricultural land use (yellow) bodes well with any other land use with the exception of industry.
- Water bodies (light blue) is compatible with any other land use with the exception of industry, with which it can nevertheless be combined diagonally.
- Industry (indigo) is compatible with urban land use in any combination and with water bodies only diagonally.

With these premises, the play is organized as follows:

(i) All 58 cubes are mixed together and then randomly allocated to the two players (29 cubes each).
(ii) The first player places one of the 29 cubes on the board.
(iii) The second player puts one of the 29 cubes with one side attached to the first player's cube, by using a color that is compatible to that of the cube that the first player placed.
(iv) Next plays the first player again, and so on, by alternating turns, until they have both placed all their 29 cubes on the board. Any cube can be placed adjacent to any other cube, anywhere on the board, so long as its color is compatible to the adjacent one(s), either horizontally, or vertically, or diagonally (in case there are restrictions to diagonal color compatibility).
(v) Every time a quartet is created (either as a linear series of four cubes or as a 2x2 square arrangement), a point is won by the player who placed the last cube that completed the quartet and two points are won for each quintet. No arrangements longer than 7 cubes in a row are allowed (this rule forces players to avoid endless linear arrangements and make 2x2 squares instead).
(vi) When all 58 cubes have been played out, the winner is the player who has accumulated the higher number of points in the course of the game.

12.3 VR/AR and Topology in Education

As stated earlier in this book, spatial dimension is one of the fundamental concepts of topology. To date however, 2D maps are used in the overwhelming majority of classes of geographical education worldwide. A school tutor tediously showing where mountains and valleys are on a 2D map is a very common experience from one's school years. This is because anaglyph (3D) maps are seldom used in teaching, except for special circumstances, to the extent that, whenever maps are demonstrated in geographical education, it is automatically implied that they are planar, which brings about a "*planification*" of geographical education.

A typical case in which using 3D representations in geographical education are more preferable than traditional 2D maps is when students learn to use relief maps in order to explain particular geographical patterns and processes. For instance, students can explain soil erosion easier and can interpret where (and why) some water flow changes direction throughout the anaglyph. Further, they can also explain the allocation of land use patterns over the geographical space much easier and may even predict what kind of land use should be encountered within a range of altitudes. This helps them formulate causal explanations linking elevation, climate and vegetation: a 3D view of the Alps extending along a West-to-East direction may lead students to anticipate that the Po valley in northern Italy can be protected from the cold northern winds and thus enjoy a more temperate climate. Similarly, students may infer which transport networks seem more feasible to build by only checking at the passes on the anaglyph map (which is certainly easier to perceive from its 3D physical or digital representation). Additionally, they may derive assessments of the human impact on a landscape by considering possible constructions such as bridges, subterranean tunnels, dams and artificial lakes. In this way, university courses of applied geography and applied geomorphology can be taught in a more natural way that would be more conducive to sharpening geographical and environmental problem solving skills (Wright et al., 2010). Evidently, these extend from physical to digital 3D maps also, although the former offer the additional advantage of involving one more sense (touch) in the educational process.

Although this planification is by no means restricted to school education only, university students of geography and related disciplines (i.e. geology) learn how to use aerial photographs and satellite imagery in order to study the earth's relief, how to draw iso-lines, watersheds etc. Despite the very many applications in geographical education, the use of 3D maps for education has hitherto been rather disregarded but current AR and VR technologies facilitate the perception of a landscape terrain in its heights and depths. Indeed, the progress in geoinformatics enabled us to use digital elevation models in tandem with satellite imagery and other technologies such as AR, so as to obtain spectacular 3D representations of landscapes and cities (Paris, New York etc). Furthermore, the science and art of "geo-visualization" enhances the perception of geographical objects and phenomena in 3D (DiBiase, 1990; McEachren & Taylor, 1994; Schultz et al., 2008), as well as spatial thinking in 3D (Herman et al., 2016; Marwa et al., 2021; Duarte et al., 2022). Simultaneously, the need for 3D models in education has been identified for other spatial sciences also (John Yu et al., 2010), i.e. it is widely recognized that design using 3D models fosters creativity in architectural education (Yamacli et al., 2005; Luescher, 2010), while the benefits of using VR and AR in geographical education and specifically for relief visualization have been repeatedly pointed out (Huang et al., 2012; Briskman et al., 2017; Carrera et al., 2018; Wang et al., 2019; Lv et al., 2017; Moore et al., 2020; Fan et al., 2020).

The educational value of these technologies becomes explicit when students are acquainted with the identification of elements of the physical landscape (buildings, land cover arrangements, flora etc) and associate other (i.e. historical, cultural) information about them. Once the students' VR/AR project has been completed (and, more likely, at some advanced level), they may propose specific changes to the real environment in order to conform with the ideal one that was digitally created. All these educational applications are possible because Virtual Reality (VR), Augmented Reality (AR), Extended Reality (XR) and Mixed Reality (MR) technologies offer unparalleled 3D experiences, while also integrating sound and motion (animation). For these reasons, they can be used to simulate landscapes and to teach how to "build" digital landscapes from basic constituents (soil types, plant species, roads, houses etc). With these technologies, the cyber-spatial digital world converges with the geo-spatial world, creating spaces within other spaces, and, for this reason, they have been implemented to create digital landscape representations in virtual and augmented environments and to teach landscape design (Honjo & Lim, 2001; Li et al., 2018; Edler et al., 2018, 2019). And yet, whereas the use of immersion and geovisualization in education contributes to furthering trainees' spatial abilities (Batty.M., 1997; Doyle & Dodge, 1998; Edler et al., 2018; Seo & Yoo, 2020; Keil et al., 2021; Rzeszewski & Naji, 2022; Lochhead et al., 2022; Bagher et al., 2023), even more powerful immersive experiences are offered by XR technologies (Çöltekin et al., 2020). It is not by accident that the commonest contents of such Virtual Worlds are digitally created artificial *geographical* settings, be they training environments (landscapes, planets, factories, urban landscapes) or imaginary spaces (futuristic cityscapes, random and fractal landscapes etc).

The main psychological effect of all these technologies is spatial, by allowing the user to enter a completely artificial (digital) space or by combining that space with the physical one and a word from topology is (metaphorically perhaps) being used to describe this experience: "immersion" (Wyshynski & Vincent, 1993). An easy and practical way to feel immersed in an AR environment is offered by AR location-based games (Youm et al., 2019) and immersive games (Keil et al., 2021), in which geo-spaces fuse with cyber-spaces. These games emerged in tandem with the expansion of GPS, reminding us that "homo ludens" (Huizinga, 1950) is always a preferable option in education. The first location-based AR game, "Botfighters", was launched in Sweden in 2001. In this game, players aimed at collecting virtual items from a city's streets while communicating with SMS. A similar game, adopted to the real map of Tokyo, was "Mogi Mogi". A not very different one, "CitiTag", was a wireless, multiplayer, location-based game, which was played using handheld PCs connected to a wireless network. In some other games such as "Pirates", the geo-space was superimposed onto the cyber-space and players became wanderers ("flâneurs"). The remarkably popular "Pokemon Go" that appeared in the first decade of the twenty-first century (An & Nigg, 2017; Rauschnabel et al., 2017; Anupama et al., 2019; Juhász et al., 2020; Bueno et al., 2020) was later followed by "Ingress" which offers an unparalleled AR experience of immersion and is particularly suited to advance geographical education (Chess, 2014; Davis, 2017; Adanali, 2021). From a combined topological-and-psychological perspective, immersion is strongly reminiscent of the Möbius band and the Klein bottle. Interestingly, the relevance of this topological surface as a model of immersive experience was interpreted by Rapoport (2011) with his observation that the transition from the physical to the virtual space and back again is effectuated in four phases, corresponding to pairs of digits 0 and 1: beginning from outside-outside (0–0), then moving to inside-outside (1–0), followed by outside-inside (0–1) and, eventually, to inside-inside (1–1).

In discussing Michel Henry's aesthetic theory, Smith (2006) suggested that "a recognition of the interdependence between autoaffection and heteroaffection" is necessary. This interdependence is nowhere better expressed than in the realm of cyber-space because VR offers the experience of the feeling of "I is an other", as the artist Orlan proclaimed in her self-hybridization attempts (Zylinska, 2002, p.223). Further, we should not lose sight from the fact that VR can create the extraordinary sense of touch termed "*proprioception*", which is "the feedback our body gives us about its current state, posture or position" (Lavroff, 1992; Wexelblat, 1993). The sense of proprioception in VR, a kind of re-discovery of self within a self-referential process, is strongly reminiscent of the exclamation "I'm turning inside out" that Hales (1999, p.168) noticed in one of Philip Dick's essays. Proprioception might thus evoke the topological process of "*eversion*" that is turning a surface or 3D object, e.g. a sphere, inside out (Francis et al., 1998; Sullivan, 2002; Francis & Sullivan, 2004; Francis, 2007).

References

Adanali, R. (2021). How geogames can support geographical education? *Review of International Geographical Education Online, 11*(1), 215–235.

An, J.-Y., & Nigg, C. R. (2017). The promise of an augmented reality game—*Pokémon GO*. *Annals of Translational Medicine, 5*(1), S11.

Anupama, C. G., Gobinath, V., & Vijay Patil, S. (2019). Aurora retrive – An augmented reality game using real world maps. *International Journal of Recent Technology and Engineering, 8*(2S4), 240–243.

Bagher, M., Sajjadi, P., Wallgrün, J. O., LaFemina, P., & Klippel, A. (2023). Virtual reality for geospatial education: Immersive technologies enhance sense of embodiment. *Cartography and Geographic Information Science, 50*(3), 233–248.

Batty.M. (1997). Virtual geography. *Futures, 29*(4/5), 337–352.

Biggs, N. L., Lloyd, E. K., Wilson, B., & J. (1976). *Graph theory: 1736–1936*. Oxford University Press.

Briskman, A., Chan, A. B., Dixon, C., & Palumbo, J. P. (2017). *The Implementation and Evaluation of Virtual Reality in Geographical Education*. PhD thesis. University of Hong Kong.

Bueno, S., Dolores Gallego, M., & Noyes, J. (2020). Uses and gratifications on augmented reality games: An examination of Pokémon go. *Applied Sciences, 10*, 1644.

Carrera, C. C., Perez, J. L. S., & Cantero, J. D. L. T. (2018). Teaching with AR as a tool for relief visualization: Usability and motivation study. *International Research in Geographical and Environmental Education, 27*(1), 69–84.

Cartwright, W. (2006). Exploring games and gameplay as a means of accessing and using geographical information. *Human IT, 8*(3), 28–67.

Chess, S. (2014). Augmented regionalism: Ingress as geomediated gaming narrative. *Information, Communication & Society, 17*(9), 1105–1117.

Çöltekin, A., Griffin, A. L., Slingsby, A., Robinson, A. C., Christophe, S., Rautenbach, V., et al. (2020). Geospatial information visualization and extended reality displays. In *Manual of digital earth* (pp. 229–277). Springer.

Da Silva, C. N. (2015). Interactive digital games for geography teaching and understanding geographical space. *Creative Education, 6*(07), 692.

Davis, M. (2017). Ingress in geography: Portals to academic success? *Journal of Geography, 116*(2), 89–97.

DiBiase, D. (1990). Visualisation in the earth sciences. *Earth and Mineral Science, 59*(2), 13–18.

Doyle, S., & Dodge, M. (1998). Towards virtual London: Developing a virtual internet G.I.S. In *Proceedings of the international conference of modelling geographical and environmental systems with G.I.S, v.II*. Hong Kong Convention and Exhibition Centre.

Duarte, L., Teodoro, A. C., & Gonçalves, H. (2022). Evaluation of spatial thinking ability based on exposure to geographical information systems (GIS) concepts in the context of higher education. *ISPRS International Journal of Geo-Information, 11*(8), 417.

Edler, D., Kühne, O., Jenal, C., Vetter, M., & Dickmann, F. (2018). Potenziale der Raumvisualisierung in Virtual Reality (VR) fur die sozialkonstruktivistische Landschaftsforschung. *Kartographische Nachrichten, 5*, 245–254.

Edler, D., Keil, J., Wiedenlubbert, T., Sossna, M., Kühne, O., & Dickmann, F. (2019). Immersive VR experience of redeveloped post-industrial sites: The example of "Zeche Holland"in Bochum-Wattenscheid. *KN – Journal of Cartography and Geographic Information, 69*(4), 267–284.

Fan, L., Miao, L., Sun, J., & Liu, C. (2020). Using AR and Unity 3D to support geographical phenomena simulations. In *2020 IEEE 5th international conference on signal and image processing (ICSIP)* (pp. 451–455). IEEE.

Fontaine, D. (2020). Landscape in computer games—The examples of GTA V and watch dogs 2. In D. Edler, C. Jenal, & O. Kühne (Eds.), *Modern approaches to the visualization of landscapes* (pp. 293–306). Springer VS.

Francis, G. K. (2007). *A topological Picturebook*. Springer.

Francis, G., & Sullivan, J. M. (2004). Visualizing a sphere eversion. *IEEE Transactions on Visualization and Computer Graphics, 10*(5), 509–515.

Francis, G., Sullivan, J. M., & Hartman, C. (1998). Computing sphere eversions. In H. C. Heger & K. Polthier (Eds.), *Mathematical visualization: Algorithms, applications and Numerics* (pp. 237–255). Springer.

Hales, N. K. (1999). *How we became Posthuman*. The University of Chigago Press.

Herman, L., Stachoň, Z., Stuchlík, R., Hladík, J., & Kubíček, P. (2016). Touch interaction with 3D geographical visualization on Web: Selected technological and user issues. *International Archives of Photogrammetry, Remote Sensing & Spatial Information Sciences, 42*.

Honjo, T., & Lim, E. M. (2001). Visualization of landscape by VRML system. *Landscape and Urban Planning, 55*(3), 175–183.

Huang, H., Tang, L., Li, J., & Chen, C. (2012). Simulation and visualization of forest fire growth in an integrated 3D virtual geographical environment-a preliminary study. In *2012 20th international conference on Geoinformatics* (pp. 1–6). IEEE.

Huizinga, J. (1950 [1938]) *Homo Ludens: A study of the play-element in culture*. Beacon Press.

John Yu, Z., Gee, G., Tabrizi, A., Redd, T., Torres, D., Miller, J., & Mourtos, N. (2010). Development and implementation of a 3D laser scanning course for land surveying. *Surveying and Land Information Science, 70*(1), 23–28.

Juhász, L., Novack, T., Hochmair, H. H., & Qiao, S. (2020). Cartographic vandalism in the era of location-based games—The case of open street map and Pokémon GO. *International Journal of Geo-Information, 9*, 197.

Keil, J., Edler, D., Schmitt, T., & Dickmann, F. (2021). Creating immersive virtual environments based on open geospatial data and game engines. *KN-Journal of Cartography and Geographic Information, 71*(1), 53–65.

Kim, M., & Shin, J. (2016). The pedagogical benefits of Sim City in urban geography education. *Journal of Geography, 115*(2), 39–50.

Lavroff, N. (1992). *Virtual reality playhouse: Explore artificial worlds on your PC*. The Waite Group Press.

Li, Z., Cheng, Y., & Yuan, Y. (2018). Research on the application of virtual reality technology in landscape design teaching. *Educational Sciences: Theory & Practice, 18*(5), 1400–1410.

Lochhead, I., Hedley, N., Çöltekin, A., & Fisher, B. (2022). The immersive mental rotations test: Evaluating spatial ability in virtual reality. *Frontiers in Virtual Reality, 3*, 820237.

Luescher, A. (2010). Concrete geometry: Playing with blocks. *International Journal of Arts & Design Education, 29*(1), 17–26.

Lv, Z., Li, X., & Li, W. (2017). Virtual reality geographical interactive scene semantics research for immersive geography learning. *Neurocomputing, 254*, 71–78.

Marwa, J. N., Noh, N. H. B., Ahyan, N. A. M., Jambari, H., & Pairan, M. R. (2021). Enhancing spatial ability skills in basic engineering drawing using a 3D solid model. *Review of International Geographical Education Online, 11*(4), 627–638.

McEachren, A. M., & Taylor, D. R. F. (1994). *Visualisation in modern cartography*. Pergamon.

Moore, A., Daniel, B., Leonard, G., Regenbrecht, H., Rodda, J., Baker, L., et al. (2020). Comparative usability of an augmented reality sandtable and 3D GIS for education. *International Journal of Geographical Information Science, 34*(2), 229–250.

Morawski, M., & Wolff-Seidel, S. (2023). Gaming & geography (education): A model of reflexive analysis of Space & Action in video games. *European Journal of Geography, 14*(3), 1–19.

Papadimitriou, F. (2020). Squares, cats and mazes: The art and magic of spatial complexity. In *Spatial complexity. Theory, mathematical methods and applications* (pp. 127–141). Springer.

Papadimitriou, F. (2022). An evaluation system for games related to geography and landscapes in education. In D. Edler, O. Kühne, & C. Jenal (Eds.), *The social construction of landscapes in games* (pp. 315–335). Springer.

Rapoport, D. L. (2011). On the fusion of physics and Klein bottle logic in biology, embryogenesis and evolution. *Neuro Quantology, 9*(4), 842–861.

Rauschnabel, P. A., Rossmann, A., & Tom Dieck, M. C. (2017). An adoption framework for mobile augmented reality games: The case of Pokemon go. *Computers in Human Behavior, 76*, 276–286.

Rzeszewski, M., & Naji, J. (2022). Literary placemaking and narrative immersion in extended reality virtual geographic environments. *International Journal of Digital Earth, 15*(1), 853–867.

Sandbrook, C. A. W., & Monteferri, B. (2015). Digital games and biodversity conservation. *Conservation Letters, 8*, 118–124.

Santos, L., Silva, N., Nóbrega, R., Almeida, R., & Coelho, A. (2020). An interactive application framework for natural parks using serious location-based games with augmented reality. In *Proceedings of the 15th international joint conference on computer vision, imaging and computer graphics theory and applications (VISIGRAPP 2020)* (pp. 247–254).

Schultz, R., Kerski, J., & Patterson, T. (2008). The use of virtual globes as a spatial teaching tool with suggestions for metadata standards. *Journal of Geography, 107*(1), 27–34.

Seo, D., & Yoo, B. (2020). Interoperable information model for geovisualization and interaction in XR environments. *International Journal of Geographical Information Science, 34*(7), 1323–1352.

Smith, J. H. (2006). Michel Henry's phenomenology of aesthetic experience and Husserlian intentionality. *International Journal of Philosophical Studies, 14*(2), 191–219.

Sullivan, J. M. (2002). Sphere eversions: From Smale through "the optiverse". In *Mathematics and art: Mathematical visualization in art and education* (pp. 201–212). Springer Berlin Heidelberg.

Swetnam, R. D., & Korenko, J. (2019). Can computer game landscapes target new audiences for landscape quality assessment? *Applied Geography, 113*, 102102.

Walford, R. (1981). Geography games and simulations: Learning through experience. *Journal of Geography in Higher Education, 5*(2), 113–119.

Wang, W., Wu, X., He, A., & Chen, Z. (2019). Modelling and visualizing holographic 3D geographical scenes with timely data based on the Holo lens. *ISPRS International Journal of Geo-Information, 8*(12), 539.

Wexelblat, A. (Ed.). (1993). *Virtual reality: Applications and explorations*. Academic.

Wright, D. R., Lidstone, J., & Stoltman, J. (2010). Evaluating and creating world maps-from basic principles. *International Research in Geographical and Environmental Education, 19*(1), 1–10.

Wyshynski, S., & Vincent, V. J. (1993). Full-body unencumbered immersion in virtual worlds. In A. Wexelblat (Ed.), *Virtual reality: Applications ad explorations* (pp. 123–146). Academic.

Yamacli, R., Ozen, A., & Tokman, L. Y. (2005). An experimental study in an architectural design studio: The search for three-dimensional form and aesthetics through clay. *International Journal of Art and Design Education, 24*(3), 308–314.

Youm, D.-H., Seo, S.-H., & Kim, J.-Y. (2019). Design and development methodologies of Kkongalmon, a location-based augmented reality game using mobile geographic information. *EURASIP Journal on Image and Video Processing, 1*, 2–11.

Zylinska, J. (2002). *The cyborg experiments*. Continuum.

Chapter 13
Geo-topology and Epistemology

Every cubic inch of space is a miracle,
every square yard of the surface of the earth is spread with the same,
every foot of the interior swarms with the same.
(Walt Whitman, 1819–1892, "Miracles")

13.1 Dimensions, Topology and Geo-topology

Topological approaches to the perception of geographical space have been adopted within the context of the philosophies of e.g. Aristotle (Levin, 2000), Wittgenstein (Peters, 2008), Whitehead (Da Costa et al., 1997), Kant (Hoefer, 2000), Prior (Simons, 2006) and Heidegger (Sean, 2009) while the urban theorist Virilio coined the term "teletopology" (Virilio, 1994, p.7) and the science fiction author Stanislav Lem proposed a "toposophy" (Broderick, 2001, p.324).

Nevertheless, the fact that non-integer dimensions attracted so much interest by geographers does not imply that integer dimensions may not present surprising properties as well and this is particularly relevant in what concerns the intersections of topology and geography with respect to the differential topology of manifolds in three and four dimensions.

Ever since Hagerstrand drew the attention of geographers to his space-time diagrams (Hagerstrand, 1982), part of the literature of human geography has been repeatedly devoted to the visualization of space-time in geographical contexts and to the perception of place-times and, besides Hagerstrand, spatial and temporal aspects of being fused together in Bakhtin's theory of "chronotopes" (Bakhtin, 1981; Holquist, 1990; Holloway & Kneale, 2000; Müller, 2010, 2016). But before expanding further into philosophy, it is important to consider spacetime as described in physics firstly and topology next.

In physics, the relationships between matter, energy and 4D space-time are described by Einstein's field equation:

$$R_{\mu\nu} - \frac{1}{2}g_{\mu\nu}R = -GT\mu\nu$$

© The Author(s), under exclusive license to Springer Nature Switzerland AG 2023

F. Papadimitriou, *Geo-Topology*, GeoJournal Library 133,
https://doi.org/10.1007/978-3-031-48185-7_13

To understand this equation, it has to be recalled how Riemannian curvature is calculated in a n-dimensional space has to be recalled: a n-dimensional geodesic surface is represented by a second order antisymmetric tensor $\sigma^{\mu\upsilon}$ and the Riemannian curvature of this surface is

$$K_{\mu\upsilon} = \frac{(\alpha\beta, \gamma\delta)\sigma^{\alpha\beta}\sigma^{\gamma\delta}}{\left(g_{\alpha\gamma}g_{\beta\delta} - g_{\alpha\delta}g_{\beta\gamma}\right)\sigma^{\alpha\beta}\sigma^{\gamma\delta}}$$

and, following "Schur's theorem", the isotropy of the Riemannian space in all its points implies its homogeneity. On the basis of this theorem, it follows that the Robertson and Walker metric is the only one compatible with a large-scale isotropic and homogeneous spatiotemporal continuum:

$$ds^2 = dt^2 - \frac{[R(t)]^2\left[(dx_1)^2 + (dx_2)^2 + (dx_3)^2\right]}{\left\{\left[1 + \frac{1}{4}K\left[(x_1)^2 + (x_2)^2 + (x_3)^2\right]\right]\right\}^2}$$

which implies that when $K = 0$ the space is Euclidean, with $K = -1$ it is hyperbolic and when $K = 1$ it is spherical. In very broad terms, this is what physics informs us about the space and the 4D space-time. Progress in differential topology in the second half of the twentieth century however, proved that the geometries of even "ordinary" 3D spaces can be significantly more difficult to understand than what was thought of before.

A number of "mysteries" relate to the topology of both 3D and 4D spaces, which might prompt for some caution before deriving geographical interpretations of the 3D space (part of which is the geographical space) with respect to the 4D space. For instance, Felix Hausdorff proved in 1914 that it is possible to divide *one* sphere into a finite set of non-measurable parts which, after re-unification, would make *two* spheres with each one of them of area equal to that of the original sphere. Although there is no physical method to do this, it is nevertheless a mathematical truth along with a further corroboration that came a decade later, when Stefan Banach and Alfred Tarski proved an equivalent theorem for volumes, the *Banach-Tarski paradox*.

With the progress of research in the differential topology of surfaces in the course of the 19th century, it was discovered (as we already know from Dehn's theorem) that every closed compact and connected surface is topologically equivalent to either orientable spheres (with a finite number of handles) or to non-orientable surfaces (such as the projective plain and the Klein bottle). Yet, none could have imagined that the three geometries (euclidean, hyperbolic, spherical) would *not* suffice for the description of 3D manifolds until 1904, when Henri Poincaré posed a famous problem on the topology of the hypersphere (the 4D equivalent of a sphere). Although a sphere is a 3D object, its surface is 2D, so a hypersphere is a 4D object of which the surface is 3D and the "*Poincaré conjecture*" *was that* "the hypersphere is the only closed and orientable 3D surface of which the fundamental group is trivial". Otherwise stated, *he* conjectured that the topological characterization of the hypersphere is equivalent to that of a sphere. This led to Thurston's "Geometrization Conjecture" which, after its proof by Perelman (Perelmann, 2002; 2003a, b), became a theorem. Following this conjecture (now theorem), we know that besides the three widely known geometries (euclidean, spherical, hyperbolic), manifolds in the 3D space may obey anyone of another five (and as yet poorly explored) geometries. Intricately related to Thurston's theorem is the "time-derivative" of any geometric structure, which was specified by Richard Hamilton and Grisha Perelman, by making use of the properties of the "heat transfer equation" that describes the gradual attenuation of heat over time and space. Perelman proved that the spatial regions necessary for Thurston's geometrization conjecture can have their points increasingly converging, while the points of distinct regions can move apart. By proving Thurston's conjecture in this way, the theorem asserts (Thurston, 1982, 1997; Scott, 1983) that every smooth 3D manifold can be cut to 2D spheres and 1-holed tori, in a way that each one of the resulting parts may have one of eight possible geometries: euclidean, spherical, hyperbolic, $H^2 \times R$, $S^2 \times R$, the geometry of the universal cover of $SL(2,R)$, the nil and the sol geometries.

Despite the classification of all differential manifolds with dimension higher or equal to five by Sergei Novikov in 1962, the problem of classification of 3D and 4D manifolds remained unsolved for a long time. With the earlier discovery of "exotic spheres" in which calculus does not apply as we usually expect it, the remarkable breakthrough by Gompf (1985) revealed the stark truth that there is an infinity of "exotic" structures in the 4D Euclidean space and, as a matter of fact, the number of these structures is "more than infinite" (it has the cardinality of the continuum). Thus, 3D and 4D spaces (the latter should not be confused with 4D space-time) present surprising geometrical and/or topological properties of which the significance for geography and cartography remains hitherto completely unknown.

However, aside of the surprising or even counter-intuitive results that topology may present to geographers (Papadimitriou 2020e, f), it has to be stressed that the topological differentiation of the geographical space is one of the fundamental mathematical premises that contribute to the production of spatial complexity in the geographical space (Papadimitriou 2020a, b, c), as well as for the perception of that complexity (Papadimitriou 2020d, e).

13.2 Topology Vs. Geography

Nonetheless, since all what geography examines corresponds to some topology and takes place in some topological space, notions of topology can help to access the very intrinsic spatial structures that govern innumerably many geographic processes as they gradually unfold from "topological eggs" to "Euclidean organisms" (De Landa, 2002, p.58). In the course of these unfoldings, topological thinking may lead us to identify some ideas that might affect the epistemology of geography. The following themes are characteristic in this respect:

(a) Two centuries ago, Cauchy proved the equivalence between the integral of a function and the area "contained" in between the x and y axes. Ever since, mathematicians used the "Riemann integral" for ordinary calculations of surfaces and volumes, so long as any discontinuities of these surfaces did not constitute a "nonmeasurable set". Yet, Henri Lebesgue revolutionized calculus in 1902, by introducing the Lebesgue integral and proving that all Riemann-integrable surfaces are Lebesgue-integrable also. The difference between the two integrals is that the Lebesgue integral subdivides the range of the function instead of subdividing the interval on which the function is defined. This difference is significant when the function has wild oscillations or discontinuities. Consequently, a geographical area with such characteristics can be calculated as *Lebesgue-integrable* rather than Riemann-integrable. This mathematical subtlety might be worth consideration in certain circumstances of geographical analysis (because, besides measuring areas, the Lebesgue integral is also useful in the study of probability distributions over a surface which are often examined in spatial analysis).
(b) As Georg Cantor showed in 1874, spaces of different dimensions (plane, line etc) can have the same number of points. This remarkable result had the unexpected repercussion that such spaces are indistinguishable at the set-theoretic level. Later, Luitzen Brouwer proved that the only way to distinguish among such omnipotent sets was by using the concept of "*connectedness*" of topological spaces: differences in connectedness enable us to distinguish between a plane and a line. A line has as many points as a plane and thus, set-theoretically, it is impossible to distinguish it from a plane. Brouwer reasoned that the way to distinguish the connectedness of a line from that of a plane is quite simple from a topological point of view. This is because, if one point is subtracted from a line, it gets separated in two lines and hence disconnected. But the same does not apply to the plane, since the plane does remain connected if one point is subtracted from it. The epistemological consequence for spatial analysis is that while geographers would have thought that polygonal spatial elements are more complex than linear elements on a map, they are in fact studying geometric elements that correspond to sets of equal cardinality in terms of topology.
(c) The study of geodesics formally begun with the philosopher Nicolas Oresme in the Middle Ages, evolved with the non-Euclidean geometries of the eighteenth century and stumbled for a while at "Hilbert's 4th problem" (until it was tackled in 1917 by Tullio Levi-Civita). The problem with describing linear features on the surface of the planet is because the sphere is locally diffeomorphic to the plane, but not locally isometric to it. A major part of cartographic and geodesic research has successfully dealt with this problem, but there are twists in the turn, lurking in an underlying topological simplification related to curves on the surface of the earth that relates to "contractions" and "*homotopy*".

Formally, let f,g: $X \rightarrow Y$ be two continuous functions. If there exists a continuous function F such that $F(x,0) = f(x)$ and $F(x,1) = g(x)$, then the function F is called a homotopy between *f* and *g*. For a fixed value *t* in I=[0,1], the function $F|_{X \times \{t\}}$ is a map from *X* to *Y*. With the values of *t* varying over the interval I, the $F|_{X \times \{t\}}$ varies continuously over a one-parameter family of continuous maps, starting from the interval I and ending with *g*. The homotopy relation is an equivalence relation on the set of all continuous functions $f: X \rightarrow Y$ (it is reflexive, symmetric, and transitive). A *path homotopy* from *f* to *g* is a continuous function f: I × I → Y such that $F(x,0) = f(x)$, $F(x,1) = g(x)$, $F(0,t) = y_0$, and $F(1,t) = y_1$ and it is an equivalence relation on the set of all paths in *Y* which have the same initial point and the same endpoint. A surface on which every loop (simple closed curve) can be contracted to a point is simply connected. In this way, the order of connectivity of the sphere is 0 and it is 4 for the pretzel surface. For a spherical surface it is $2p$ if that surface has p handles (if $p = 4$, then there are 8 pairs of holes connected by those 4 handles). All transformation groups form a "fundamental group" (with the algebraic sense of the term "group") and any topological (rubber-like) transformation (no tearing apart, no cutting, no discontinuous changes allowed) of a line is said to be "homotopic" to the original line. The fundamental group of the sphere is trivial because any closed path on the sphere can be contracted to a point (Fig. 13.1). If the sphere has one handle, it is non-trivial, because there is (at least) one path which can not be contracted to a point. So the fundamental group of surfaces differentiates them according to their *homotopy*.

The concept of homotopic paths demonstrates the epistemological difference between topology and geography. For topology, all lines on the surface of the earth (which is homeomorphic to a sphere) that may be studied by a geographer (e.g. railway connections, road connections) are homotopy-equivalent. This is because the surface of the planet (it does not matter for topology that the earth is not a perfect sphere) is an all–encompassing space for all its linear elements, which can eventually be contracted to a topological equivalence of one another. Otherwise put, thinking "homotopically" about the 3D surface of the planet, it is from the point of view of topology "uninteresting" to examine different curves on the surface of the earth (whether natural or human-made), since they are all homotopic. But from the

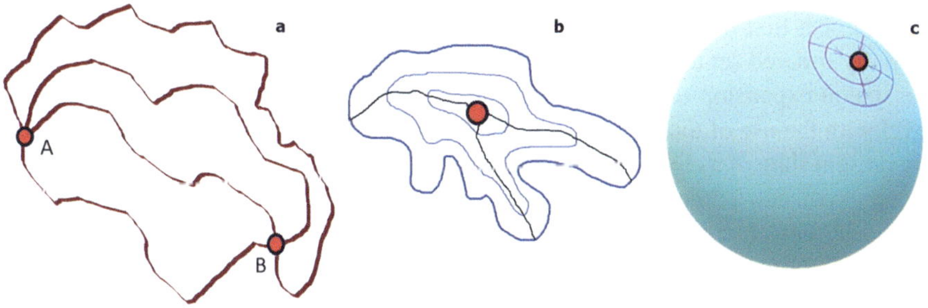

Fig. 13.1 Four different homotopic paths from point A to point B (**a**). Every simple closed curve on a simply connected surface can be contracted to a point. Here are shown possible contractions on the plane (**b**) and on the ball (**c**)

geographic point of view, every single linear element on the face of the earth is considered as a geometric entity associated with its own particular qualities (cultural background, history, legends etc) and its spatial proximity with other elements of the geographical space. This point highlights the significance of geography, for it is the geographer's role to examine the relational, historical, semantic etc. differences of otherwise homotopic spatial features.

(d) With the exception of the works aiming to derive set-theoretic descriptions of the geographic space (Beguin & Thisse, 1979) and to understand the way we perceive a geographical space or place (Desbarats, 1976), the interface of topology and logic has been a rather poorly researched field in geography. By adopting notions shared by topology or logic such as Borel algebras, Lebesgue measures, disjoint bounded subsets etc. (Egenhofer et al., 1989), the interest in topological constructions sparked in the late 1990s with the shift of geographers' interest towards GIS topologies concurrently with research in spatial databases that subsequently led to the use of linguistic variables for the description of the geographical space. Furthering those approaches that used "classic" mathematics however, constructivism and intuitionism might bring quite different approaches by using notions of "Kripke models" and "topos theory". The common ground upon which both classical and constructivist approaches are based upon is the consideration that, topologically, a well-defined area of the geographical space is a bounded metric space with finite cardinality. Of particular interest to geography can be the concept of "locatedness" which is a function of separatedness and inclusiveness in space, giving birth to "place" and "locality" with separatrices, defining the "lebenstraum" of the locus. Locatedness resonates "the feeling of "appropriated space, conceived space, space grasped through the individual's ceaseless activity to locate properly within the geographical space" so that "when the subject is able to tread the maze without error or with only rare errors, the whole maze becomes the locality with appropriate movements" (Tuan, 1977, p.72).

"Topos" models are generalizations of Kripke models. Topos is a category (with the mathematical sense of the word) that behaves like the category of sets: any partially ordered set can become a category, if a unique morphism is applied. Then, a Kripke model for a first-order spatial language L may be thought of as a functor from a partially-ordered set to the set of classical models of L. The geographic processes of the spatial expansion of an urban sprawl for instance, can be described at a general level with the use of topos models. Formalizing the process of urban growth, if S_1 is the present state of a city and S_2 its next (future) state, the morphism m_1 describes the urban expansion process from the state S_1 to the state S_2 and the urban cycle may close up with another morphism, m_2, that describes the retraction phase of the urbanization cycle. Kripke models might be viewed as a postmodernist's view of geographical processes, since postmodernism "insists that we cannot aspire to any unified representation of the world or picture it as a totality full of connections and differentiations rather than as perpetually shifting fragments" (Harvey, 1989, p.52). The cinema-like perception of these shifting fragments therefore conditions any "truth" about any "place" as a first state of a Kripke model, whereas any subsequent state may have its own "truth", which, in this sense, is equivalent to saying that it will be associated with its own narrative, with its own set of explanations and interpretations.

(e) Brouwer's fixed point theorem asserts that every continuous mapping f from B^n to B^n has a fixed point for some point x in B^n: $f(x) = x$. Additionally, the *Borsuk-Ulam Theorem* (Lovacz & Schrijver, 1998) introduces a symmetry to the fixed-point property, since it concerns a mapping from x to $-x$: for every continuous mapping f from S^n to R^n there exists a point x in S^n such that $f(x) = f(-x)$. If, for instance, the surface of a rubber ball is crumbled to make it completely flat, then two antipodal points of the ball will coincide after the ball was laid flat.

Perhaps, the relevance of this property of antipodal points to geographical thinking becomes more explicit by considering a similar theorem, the *Lyusternik-Shnirel'man Theorem*, which asserts that for any cover $\{V_1, V_2, \ldots, V_{n+1}\}$ of the sphere S^n by $n + 1$ closed sets, there is at least one set containing a pair of antipodal points. This statement is tantamount to claiming that the intersection of V_i with $-V_i$ is non-void. Hence, while studying landscape change from time t_1 to time t_2, statements on "change" are valid for the entire area studied *with the exception of one point* always, which has remained unchanged. These theorems therefore might lead us to cast doubts on the certainty that a landscape changed *completely* from time t_1 to time t_2 (since there always remains one point that did not change). From a philosophical perspective, the point that remains invariable throughout time despite any landscape transformations and even if it can not be identified in the physical space, is reminiscent of the "genius loci" The "spirit of place" has been a recurrent theme in human geography (Strecker, 2000; Smil, 2001; Barnes, 2004; Bidwell & Browning, 2010; Christou et al., 2019; Vecco, 2020; Andalucia et al., 2023; De Silva, 2023; Rana, 2023). Should geography accept that a point of the geographical space (a dimensionless entity) inevitably remains inaccessible to human knowledge? Expectedly, this (potenitally inconvenient) question may open up ample room for explanations and interpretations in the future.

References

Andalucia, A., Ginting, N., Aulia, D. N., & Hadinugroho, D. L. (2023). Spirit of place as an attraction of heritage area in Medan City, Indonesia. *Environment-Behaviour Proceedings Journal, 8*(23), 287–293.

Bakhtin, M.M. (1981). *The dialogic imagination: Four essays by M.M.* (C. Emerson & M. Holquist, Trans.). University of Texas Press.

Barnes, T. J. (2004). Placing ideas: Genius loci, heterotopia and geography's quantitative revolution. *Progress in Human Geography, 28*(5), 565–595.

Beguin, H., & Thisse, J. F. (1979). An axiomatic approach to geographic space. *Geographical Analysis, 11*(4), 325–341.

Bidwell, N. J., & Browning, D. (2010). Pursuing genius loci: Interaction design and natural places. *Personal and Ubiquitous Computing, 14*, 15–30.
Broderick, D. (2001). *The spike*. Forge.
Christou, P. A., Farmaki, A., Saveriades, A., & Spanou, E. (2019). The "genius loci" of places that experience intense tourism development. *Tourism Management Perspectives, 30*, 19–32.
Da Costa, N. C. A., Bueno, O., & French, S. (1997). Suppes predicates for space-time. *Synthese, 112*(2), 271–279.
De Landa, M. (2002). *Intensive science and virtual philosophy*. Bloomsbury.
De Silva, W. (2023). Comprehending genius loci, towards spiritual sustainability: Lessons from Buddhist heritage city Anuradhapura. *International Journal of Heritage Studies, 29*(1–2), 1–20.
Desbarats, J. (1976). Semantic structure and perceived environment. *Geographical Analysis, 8*, 453–467.
Egenhofer, M. J., Frank, A. U., & Jackson, J. P. (1989). A topological data model for spatial databases. In A. P. Buchmann, O. Günther, T. R. Smith, & Y. F. Wang (Eds.), *Design and implementation of large spatial databases. SSD 1989* (Lecture notes in computer science) (Vol. 409, pp. 271–286). Springer.
Gompf, R. E. (1985). An infinite set of exotic R4's. *Journal of Differential Geometry, 21*, 283–300.
Hagerstrand, T. (1982). Diorama, path, project. *Tijdschrift voor Economische en Sociale Geografie, 73*(6), 323–329.
Harvey, D. (1989). *The condition of postmodernity*. Oxford/Blackwell.
Hoefer, C. (2000). Kant's hands and Earman's pions: Chirality arguments for substantial space. *International Studies in the Philosophy of Science, 14*(3), 237–256.
Holloway, J., & Kneale, J. (2000). Mikhail Bakhtin. In M. Crang & N. Thrift (Eds.), *Thinking space* (pp. 71–88). Routledge.
Holquist, M. (1990). *Dialogism: Bakhtin and his world*. Routledge.
Levin, S. B. (2000). *The ancient quarrel between philosophy and poetry revisited: Plato and the Greek literary tradition*. Oxford University Press.
Lovacz, L., & Schrijver, A. (1998). A Borsuk theorem for antipodal links and a spectral characterization of linklessly embeddable graphs. *Proceedings of the American Mathematical Society, 126*(5), 1275–1285.
Müller, T. (2010). Notes toward an ecological conception of Bakhtin's chronotope. *Ecozon@: European Journal of Literature, Culture and Environment, 1*(1), 98–102.
Müller, T. (2016). The ecology of literary Chronotopes. In *Handbook of ecocriticism and cultural ecology* (pp. 590–604). De Gruyter.
Papadimitriou, F. (2020a). *Spatial complexity. Theory, mathematical methods and applications*. Springer.
Papadimitriou, F. (2020b). The topological basis of spatial complexity. In *Spatial complexity. Theory, mathematical methods and applications* (pp. 63–79). Springer.
Papadimitriou, F. (2020c). Geophilosophy and epistemology of spatial complexity. In *Spatial complexity. Theory, mathematical methods and applications* (pp. 263–278). Springer.
Papadimitriou, F. (2020d). Spatial complexity, psychology and qualitative complexity. In *Spatial complexity. Theory, mathematical methods and applications* (pp. 229–242). Springer.
Papadimitriou, F. (2020e). Exploring spatial complexity in 3D. In *Spatial complexity. Theory, mathematical methods and applications* (pp. 101–113). Springer.
Papadimitriou, F. (2020f). Spatial complexity in 4-and-higher dimensional spaces. In *Spatial complexity. Theory, mathematical methods and applications* (pp. 115–123). Springer.
Perelmann, G. (2002, November 11). The entropy formula for the Ricci flow and its geometric applications. arXiv:math.DG/0211159 v1.
Perelmann, G. (2003a). Ricci flow with surgery on three manifolds. arXiv:math.DG/0303109 v1 March 10, 2003, Preprint.
Perelmann, G. (2003b). Finite extinction time to the solutions to the Ricci flow on certain three manifolds. arXiv:math.DG/0307245 July 17, 2003, preprint.

Peters, M. A. (2008). Wittgenstein as exile: A philosophical topography. *Educational Philosophy and Theory, 40*(5), 591–605.

Rana, J. R. (2023). Pilgrimage archetype and cultural heritage of Jagannath temple in Varanasi, India. *Esempi di Architettura, International Journal of Architecture and Engineering, 10*(2), 270–283.

Scott, P. (1983). The geometries of 3-manifolds. *Bulletin of the London Mathematical Society, 15*, 401–487.

Sean, R. (2009). Heidegger's topology: Being, place, world. *Australasian Journal of Philosophy, 87*(1), 169–171.

Simons, P. (2006). The logic of location. *Synthese, 150*(3), 443–458.

Smil, V. (2001). Genius loci. *Nature, 409*(6816), 21–21.

Strecker, I. (2000). The "genius loci" of Hamar. *Northeast African Studies, 7*(3), 85–118.

Thurston, W. P. (1982). Three dimensional manifolds, Kleinian groups and hyperbolic geometry. *Bulletin of the American Mathematical Society, 6*, 357–381.

Thurston, W. P. (1997). In S. Levy (Ed.), *Three-dimensional geometry and topology* (Vol. 1). Princeton University Press.

Tuan, Y.-F. (1977). *Space and place: The perspective of experience*. University of Minnesota Press.

Vecco, M. (2020). Genius loci as a meta-concept. *Journal of Cultural Heritage, 41*, 225–231.

Virilio, P. (1994). *The vision machine*. British Film Institute.

Chapter 14
Personal Geo-topologies

Love is space and time measured by the heart.
"L'amour c'est l'espace et le temps rendus sensibles au cœur"
(Marcel Proust, 1871–1922, "La Prisonnière")

14.1 Topological Daily Routines in the Twenty-First Century

One way to appreciate the increasing "topologisation" of everyday life is by identifying the notions of topology that are encountered in usual routines of a working day. So this is how a "topological day" might go in the early twenty-first century:

> It's 8 am in the morning; time to get up.
> My shirt needs ironing. But ironing out the shirt's crumples restores its topological equivalence to the completely flat shirt: here enters topology with "Brouwer's fixed point theorem".

Place two planar objects of equal size one atop of the other. If one of them (say the upper one) were deformed (without tearing) and thrown over the lower one (which remained intact), then there will always be at least one point of the deformed object that has a one-to-one correspondence with its corresponding point (in terms of coordinates) with the lower one below (Fig. 14.1).

I open my cupboard to choose among my jackets, pullovers and sweaters. I realize that each garment has its own "genus" (number of holes). Trousers for instance, have a different genus than sweaters. Pants correspond to what is called a "cobordism" in topology: a fusion of borders of two different surfaces, while a pullover has a different topology than a jacket (Fig. 14.2).

But the topology of garments is not about genus only, for it's about braids too. The way that some knitwear, cloth or garment is woven depends on the braiding scheme that was used to fabricate it. Braids are sets of strands joining two planes in the 3D space. One or more of the strands may be atop of the other. A braid of four strands for instance, b_i, may appear in different forms:

$$b_1, b_1^{-1}, b_2, b_2^{-1}, b_3, b_3^{-1}$$

© The Author(s), under exclusive license to Springer Nature Switzerland AG 2023

F. Papadimitriou, *Geo-Topology*, GeoJournal Library 133,
https://doi.org/10.1007/978-3-031-48185-7_14

Fig. 14.1 Following Brouwer's theorem, after folding (without tearing) a sheet of A4 paper and holding it atop of another that lies stretched beneath, it is ensured that there is at least one point on the folded sheet that is at an one-to-one correspondence to a point on the sheet beneath

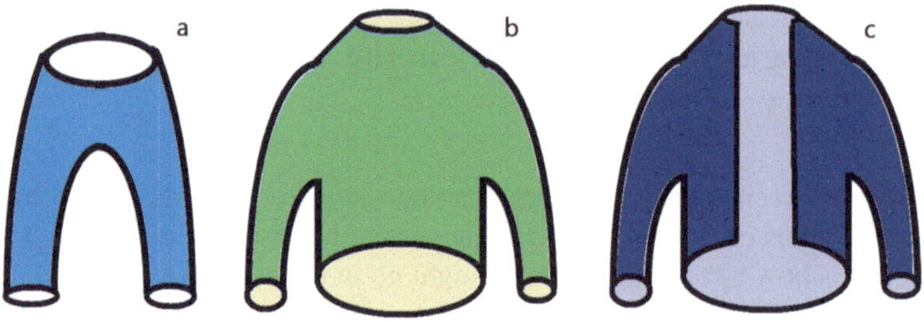

Fig. 14.2 A pair of pants (**a**) is topologically equivalent to the topology of a "cobordism". Topologically, a pullover (**b**) is inequivalent to a jacket (**c**)

depending on whether one of its two strands is atop of the other (Fig. 14.3) and if all strands are disentangled, then the braid is a "neutral braid" (e).

Having ironed my shirt, I look for my favorite tie now, which I will tie up with a "Windsor knot" (Fig. 14.4) and I realize the topological background here: knots are objects of research in topology and this particular knot is topologically equivalent (Fink & Mao, 1999) to the simple "trefoil" knot.

Although knots can be distinguished from one another on the basis of their qualitative properties, characteristic polynomials are specific to each one of them. For instance, the trefoil knot can be created by means of a set of three parametric equations:

$$x = \sin t + 2\sin 2t$$
$$y = \cos t - 2\cos 2t$$
$$z = -\sin 3t$$

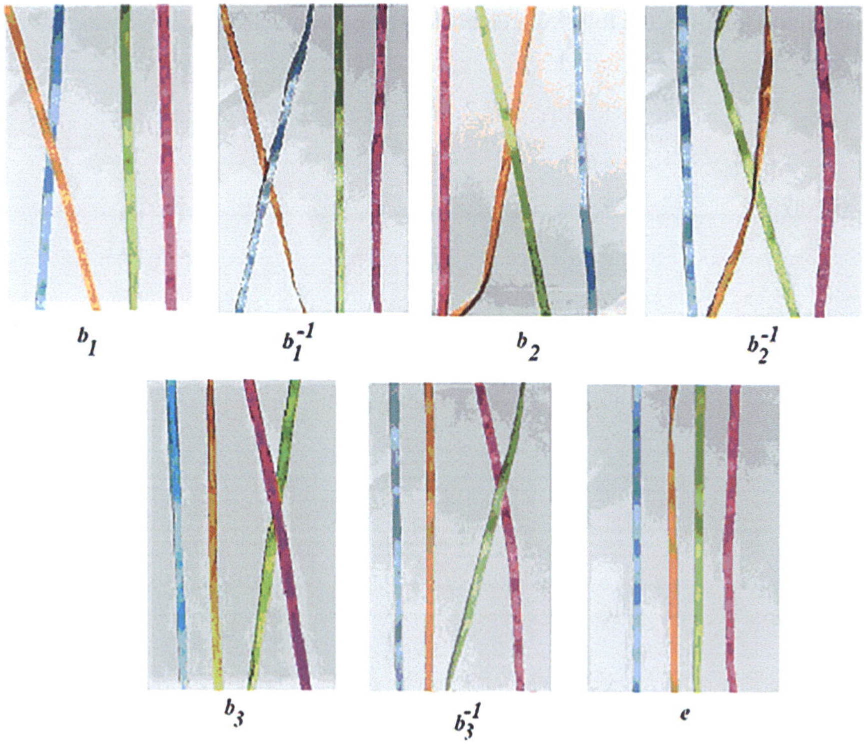

Fig. 14.3 Topologically inequivalent forms of braids that consist in four strands

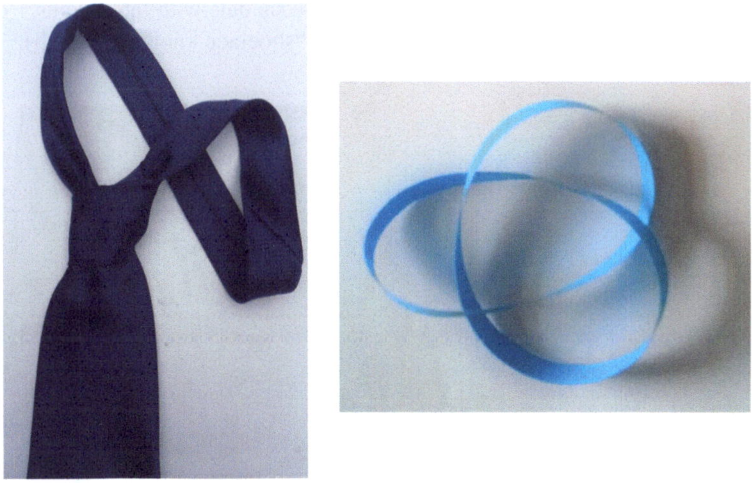

Fig. 14.4 Tie knots can be viewed through the lenses of knot theory. The "Windsor knot" of a tie is "topologically homeomorphic" to a simpler knot, the "trefoil"

The trefoil knot has the following characteristic polynomials:
Alexander:

$$x + \frac{1}{x} - 1$$

Conway:

$$x^2 + 1$$

Homfly:

$$-x^4 + x^2y^2 + 2x^2$$

Jones:

$$-\frac{1}{x^4} + \frac{1}{x^3} + \frac{1}{x}$$

Kauffman:

$$x^5y + x^4y^2 - x^4 + x^3y + x^2y^2 - 2x^2$$

Shoes are next: I have "Oxford mens' shoes" and other design forms (i.e. the "Buck", the "Blucher", the "Derby" and the "Balmoral"), which are all of the same genus (zero). I must get dressed up for I have to go to a multinational company, so I will definitely need zero-genus shoes (not the sandals with their genus 1 or the flip flops with genus 2) and I tie them up with a "shoelace knot" (Fig. 14.5).

Fig. 14.5 A "sholelace knot"

Fig. 14.6 The "Borromean rings" are three interlocked rings

Nonetheless, I notice that women's shoes have a wider range of genus values than my own (men's) shoes: the genus of their shoes ranges from zero (heels) to at least four (i.e. espadrille). They also wear earrings that offer the opportunity to increase the genus of their attire, as well as singlets, anklets, armlet band rings (genus 1), coiled bracelets, even triple rings such as Borromean rings (Fig. 14.6), which can be described by a "braid word":

$$b_1^{-1} b_2 b_1^{-1} b_2 b_1^{-1} b_2$$

As soon as I get dressed with topologically inequivalent garments and step out of my home, I realise that the urban environment is replete with topology-based and topology-driven networks, to the extent that, indeed, "whenever we look at life, we look at network" (Capra, 1996, p.82).

Upon switching my mobile phone on, I recall that the grid-based topological models of mobile telephony are typically based on hexagons but the growing demand for increased connectivity in cellular telecommunication networks led to the exploration of alternative topologies, such as Voronoi polygons (Fig. 14.7a) instead of hexagons (Fig. 14.7b) that are suitable for the "topology of heterogeneous networks" such as the "Hetnet" (Maksymyuk et al., 2015; Fadoul, 2020).

Walking down the streets towards the city centre and recalling the problem of Königsberg bridges that marked the beginning of topology, I realise it's not very distant from one of the modern fields of geography which is no other than transport geography. It brings to mind one of the most important theoretical premises of the topological analysis of street networks (see, e.g. Jiang & Claramunt, 2004; Jiang & Liu, 2009; Jiang et al., 2014; Jiang, 2015).

This is that lengths, distances, orientations and other geometric characteristics of streets do not matter in a topological analysis. Instead, topology focuses on

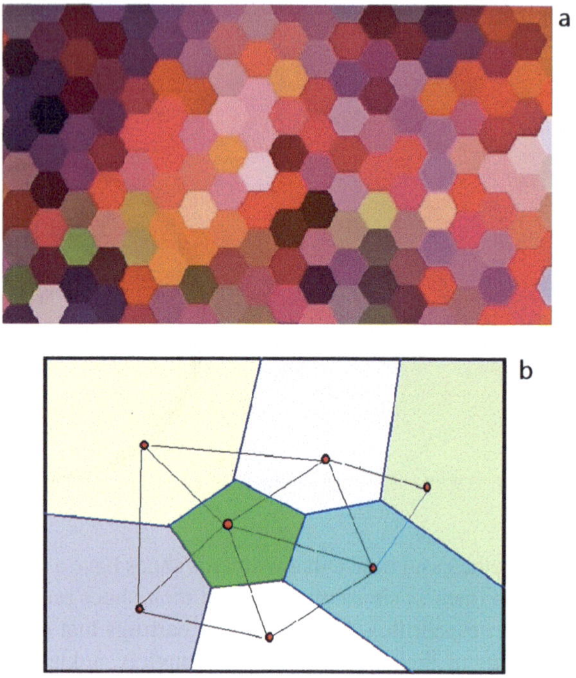

Fig. 14.7 Hexagonal topology for mobile telephony (**a**) and Voronoi polygons for heterogeneous networks of mobile telephony (**b**)

qualitative and relational features such as junction points. So geometrically dissimilar street networks may observe the same topological relationships (Fig. 14.8).

After my work I will visit a new friend's place so I need to plan how to get there by traveling the shortest distance; that's called "route optimization". The entire transport infrastructure that has been created in the city I live in (roads, bus routes, streets and highways) was based on some kind of topological analyses with a view to increasing efficacy (Mauz et al., 2003; Xue et al., 2010; Pavan et al., 2010), also to minimize environmental impact on adjacent green areas (Sienkiewicz & Holyst, 2005) but, moreover, for route optimization.

Bus routes and subway stations have been located by taking into account the needs of sizeable populations instead of particular individual needs, so I have to rely upon my topological skills (network analysis in this case) to determine the route that would minimize travel time. I need to consider several possible bus stops and a few subway stations, combine them together and come up with the best possible schedule. That's quite some topological thinking, I reckon. But there is work to be done before having fun with friends.

After getting on board a bus that would get me to my work, I check my mobile phone's connection to the Internet. The Internet is a network in which hundreds of thousands of html pages are being linked to other ones. Each one of them corresponds to a node of the network and each link to an edge: a vast network of unfathomable complexity. A glimpse of the ultra-high connectivity of each website

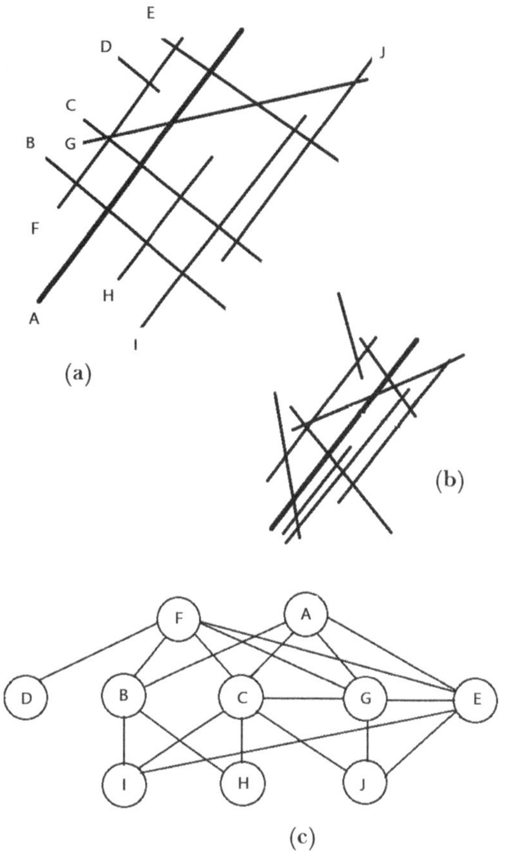

Fig. 14.8 Street networks (**a**) and (**b**) differ in street lengths and orientations, although they can be described by the same network topology which is given by a network model (**c**) that explains the structure of both (**a**) and (**b**)

can be visualised with the help of specially designed automatic indexing bots, the "web crawlers" (Fig. 14.9).

But it is not just the Internet that's replete with networks; it's the physical space also (the "geo-space"). The network of the 30 closest colleagues at my work (represented as nodes) and with unequal distribution of links among those nodes is a "small worlds" social network, with few but effective links between nodes maximizing the communication among nodes (Fig. 14.10). It is different from a "random network", in which all nodes are interconnected at random with many linkages. Small worlds networks abound in the cyber-spaces also.

The *distance* of each from every other node is given by the distance matrix (Table 14.1), a symmetric 30x30 matrix in this case:

The average shortest path length between any two nodes is 2.62 and as the distance matrix shows, the distances range between 1 and 4.

The topological term "connectivity" refers to how nodes are connected and it is measured in various ways. For instance, the "prestige" index of a node is the sum of inbound links to that node from all other nodes. In undirected graphs, the centrality index is the sum of all edges attached to a node.

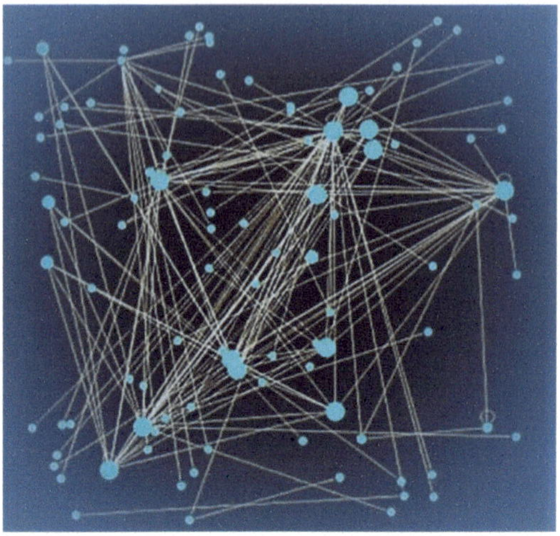

Fig. 14.9 The network of the 100 most important websites (appearing as nodes) that are linked to a meteorological site, as recorded and plotted by a "web crawler"

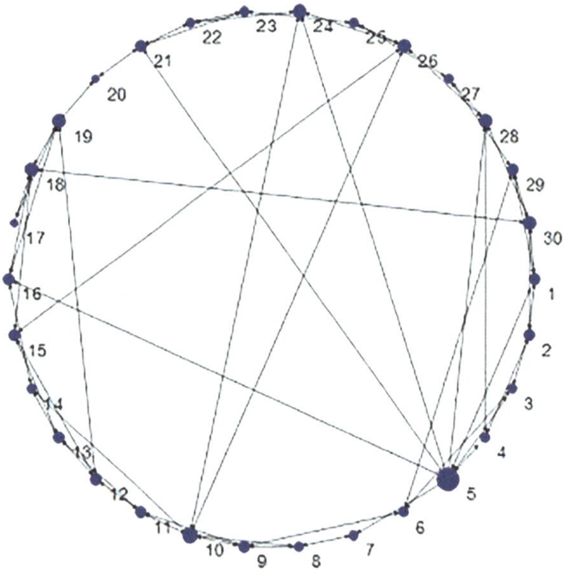

Fig. 14.10 A "small world"-type network with 30 nodes and 58 links among them: some nodes are more influential than others

The *prestige* index reflects the extent to which a person is chosen by all other ones for friendship:

$$P = \frac{choices}{g-1}$$

and takes 1 as its maximum value, where g is the number of nodes. This index is calculated on directed graphs only, displaying the directions pointing to the node i as choices.

Table 14.1 The matrix of distances from node to node in the social network

1 |	0 1 2 2 1 2 2 3 3 3 4 4 4 3 3 2 3 2 3 3 2 3 3 2 3 3 2 2 1 1
2 |	1 0 1 1 2 2 3 4 3 4 4 4 4 4 3 3 3 2 3 4 3 4 4 3 4 3 3 2 2 1
3 |	2 1 0 2 1 1 2 3 2 3 3 4 4 3 4 2 4 3 3 3 2 3 3 2 4 3 3 2 2 2
4 |	2 1 2 0 1 3 2 3 4 3 4 4 4 3 3 2 4 3 3 3 2 3 3 2 3 2 2 1 3 2
5 |	1 2 1 1 0 2 1 2 3 2 3 3 3 2 3 1 3 2 2 2 1 2 2 1 3 2 2 1 2 2
6 |	2 2 1 3 2 0 3 2 1 2 2 3 3 3 4 3 4 3 4 4 3 4 4 3 3 3 2 3 1 2
7 |	2 3 2 2 1 3 0 1 2 2 3 4 4 3 4 2 4 3 3 3 2 3 3 2 4 3 3 2 3 3
8 |	3 4 3 3 2 2 1 0 1 1 2 3 3 2 3 3 5 4 4 4 3 3 4 2 3 2 4 3 3 4
9 |	3 3 2 4 3 1 2 1 0 1 1 2 2 2 3 3 4 4 3 4 4 3 4 2 3 2 3 3 2 3
10 |	3 4 3 3 2 2 2 1 1 0 1 2 2 1 2 2 4 3 3 4 3 2 3 1 2 1 3 2 3 3
11 |	4 4 3 4 3 2 3 2 1 1 0 1 1 2 2 3 3 3 2 3 4 3 4 2 3 2 4 3 3 4
12 |	4 4 4 4 3 3 4 3 2 2 1 0 1 2 1 2 2 2 1 2 3 4 4 3 3 2 4 3 4 3
13 |	4 4 4 4 3 3 4 3 2 2 1 1 0 1 1 2 3 2 2 3 4 4 4 3 3 2 4 3 4 3
14 |	3 4 3 3 2 3 3 2 2 1 2 2 1 0 2 1 3 2 2 3 3 3 4 2 3 2 4 3 4 3
15 |	3 3 4 3 3 4 4 3 3 2 2 1 1 2 0 2 2 1 2 3 4 3 3 2 2 1 3 2 3 2
16 |	2 3 2 2 1 3 2 3 3 2 3 2 2 1 2 0 2 1 1 2 2 3 3 2 4 3 3 2 3 2
17 |	3 3 4 4 3 4 4 5 4 4 3 2 3 3 2 2 0 1 1 2 3 4 4 4 4 3 4 3 3 2
18 |	2 2 3 3 2 3 3 4 4 3 3 2 2 2 1 1 1 0 1 2 3 4 4 3 3 2 3 2 2 1
19 |	3 3 3 3 2 4 3 4 3 3 2 1 2 2 2 1 1 1 0 1 2 3 3 3 4 3 4 3 3 2
20 |	3 4 3 3 2 4 3 4 4 4 3 2 3 3 3 2 2 2 1 0 1 2 2 3 3 4 4 3 4 3
21 |	2 3 2 2 1 3 2 3 4 3 4 3 4 3 4 2 3 3 2 1 0 1 1 2 2 3 3 2 3 3
22 |	3 4 3 3 2 4 3 3 3 2 3 4 4 3 3 3 4 4 3 2 1 0 1 1 2 2 3 3 4 4
23 |	3 4 3 3 2 4 3 4 4 3 4 4 4 4 3 3 4 4 3 2 1 1 0 2 1 2 2 3 3 4
24 |	2 3 2 2 1 3 2 2 2 1 2 3 3 2 2 2 4 3 3 3 2 1 2 0 2 1 3 2 3 3
25 |	3 4 4 3 3 3 4 3 3 2 3 3 3 3 2 4 4 3 4 3 2 2 1 2 0 1 1 2 2 3
26 |	3 3 3 2 2 3 3 2 2 1 2 2 2 2 1 3 3 2 3 4 3 2 2 1 1 0 2 1 3 2
27 |	2 3 3 2 2 2 3 4 3 3 4 4 4 4 3 3 4 3 4 4 3 3 2 3 1 2 0 1 1 2
28 |	2 2 2 1 1 3 2 3 3 2 3 3 3 3 2 2 3 2 3 3 2 3 3 2 2 1 1 0 2 1
29 |	1 2 2 3 2 1 3 3 2 3 3 4 4 4 3 3 3 2 3 4 3 4 3 3 2 3 1 2 0 1
30 |	1 1 2 2 2 2 3 4 3 3 4 3 3 3 2 2 2 1 2 3 3 4 4 3 3 2 2 1 1 0

Alternatively, the *influence* of a person is measured by the number of other persons who can reach it.

The *degree prestige* (*P*) of a node is the sum of inbound connections to it, coming from all nodes that are adjacent to it divided by the number of nodes minus one.

Another measure of a person's prestige within the network's topology is the *proximity prestige* (*PP*): the ratio of the number of other persons that can reach it, divided by the average distance these other persons have from it.

Combining prestige and distances, another measure is the *influence range closeness centrality* (*IRCC*) of a node. It is the fraction of nodes reachable by each node divided by the average distance of each one of the nodes.

Table 14.2 The values of P, PP and IRCC per node of the social network

Node links	*P*	*PP*	*IRCC*	
1	4	0.13793	0.34940	0.40278
2	4	0.13793	0.30526	0.34524
3	3	0.10345	0.25337	0.38158
4	3	0.10345	0.31868	0.37662
5	8	0.27586	0.49153	0.50877
6	3	0.10345	0.23577	0.36709
7	2	0.06896	0.24246	0.36709
8	3	0.10345	0.25392	0.34524
9	4	0.13793	0.27586	0.37662
10	6	0.20690	0.40278	0.42647
11	4	0.13793	0.25676	0.36709
12	4	0.13793	0.31694	0.36250
13	4	0.13793	0.23690	0.34940
14	3	0.10345	0.32574	0.38667
15	4	0.13793	0.35802	0.39726
16	4	0.13793	0.40664	0.43939
17	2	0.06896	0.25966	0.31868
18	5	0.17241	0.39189	0.40845
19	5	0.17241	0.34761	0.39189
20	2	0.06896	0.32228	0.34524
21	4	0.13793	0.33793	0.39189
22	3	0.10345	0.32184	0.34524
23	3	0.10345	0.28713	0.33333
24	3	0.13793	0.42647	0.43939
25	3	0.10345	0.33333	0.36250
26	5	0.17241	0.42647	0.44615
27	3	0.10345	0.33721	0.35366
28	5	0.17241	0.43284	0.44615
29	4	0.13793	0.33721	0.37662
30	5	0.17241	0.39189	0.40845

In this network, node 5 has the highest prestige (0.27586), while nodes 7, 17 and 20 have the lowest (0.06896). Node 5 has the highest proximity prestige and IRCC also (Table 14.2).

Working with my colleagues in the company is pleasant, but I wonder what model describes my network of friends best. Perhaps, the topology of my friendships at my office conforms with an "ultrametric topology" (Papadimitriou, 2013), in which all triangles are isosceles and the same topology applies to friendships in the cyberspace.

I now arrive at my workplace. "TEL" (Topology in Everyday Life) is a high-tech company that has implemented networks of geosensors, IoT (Internet of Things) and IoE (Internet of Everything) at various locations of its premises (networked sensors and communicating devices, recording movements, pressure, temperature changes etc). Each one and all these networks obey their own topologies. TEL undertakes financial analyses by monitoring the WTW (the Word Trade Web), an ubiquitous network that includes interbank connections and global markets. TEL's main aim is to gain a deeper understanding of the topological properties of the WTW, which can be quite revealing of how the world economy works (Bonanno et al., 2003; Boss et al., 2004; Garlaschelli & Loffredo, 2004; Garlaschelli et al., 2005; Fagiolo et al., 2008, 2009; Serrano & Boguná, 2003; Kiyota, 2022).

And with these thoughts, my topological day has just begun.

14.2 Philosophising Personal Geo-topologies

Mingling cyber-spaces with geo-spaces together constitutes probably the most exciting facet of the ongoing topological revolution. For instance, topology can be important in deciphering the relationships between twits and the geo-space (Jiang & Ren, 2019), providing a conceptual framework for examining the spatial perception of "virtual worlds" "synthetic worlds", "virtual environments" and "digital worlds". In these artificial worlds, mixing the physical with the virtual results in a "mixed reality" (Rheingold, 2002) that becomes possible through the interpenetration of networks of physical and electronic settings. In his "Lucidity Pact", Baudrillard claimed that Reality has fallen prey to Virtual Reality (Baudrillard, 2013), a process that ends in what he calls "Integral Reality". As for geography, Cubitt (1998 p.50-53) wrote that the dream of cartography is "the endlessly updatable map" and that "satellite maps constitute the imperial encyclopedia", an endlessly updatable map in which "we are both mapped and mapping subjects of corporate knowledge systems", in a Gödelian self-reference.

Humans have always sought better environments to live in. Some called this inherent desire for a better world "utopism". Whatever the a more appropriate word about it may be, and even when we are not in position to create such environments for ourselves in the physical space (referred to as the "geo-space" here), we now have become able to design them in the cyber-space (the Internet plus the innumerable artificial digital spaces that have been created). In this way, our aspirations for better worlds has partly migrated from the geo-space to the cyber-space. Whilst it requires an architect and a well thought investment project to build a "utopistic" dwelling in the geo-space, it may only need some software and/or good programming skills to make one in the cyber-space. For this reason, virtual worlds abound on the Web; alternative "heterotic" spaces, "heterotopias", contributing to the creation of personal "autotopographies" (Gonzalez, 1995). But it is not only about "heterotopias" (at least in the sense of Foucault's theory), because we begun to model desired worlds and then turn them into reality in the cyber-space (i.e. the metaverse). What happens in the cyber-space is as "real" as what happens in the geo-space and "this is the hallmark of the age of postmodernism and simulation results in nostalgia for the real" (Bukatman, 2005, p.97). Such simulations emerge whereas "networks also overlap and connect with other networks, eventually making us realize that "it is a small world" (Urry, 2003, p.52). These trends are already discernible: we already talk about "web intelligence", "collective intelligence" and "geospatial intelligence" as we increasingly dwell not only in the physical geographical space, but in the cyber-space also. Neogeography is a characteristic example of such a fusion of the geo- with the cyber- in geography. It may be thought of seeking its origin in the practice of flâneurie, which was first mentioned in literary works of Charles Baudelaire and Edgar Alan Poe (Lauster, 2007), adopting practices of "psychogeography" (Fleischer, 2001). In his "Theory of the Derive", Guy Debord proposed "experimental derives" in which imprecise maps of influences can be drawn, depicting aspects of urban life (Debord, 1989) by a wandering that is at times structured and at times chaotic. Either ways, the old psychogeography re-emerges as neogeography in the digital era with a digital psychogeography as wandering in the cyber-space (Elias, 2010; Kvas, 2014). In both geographies, it is about emphasizing personal and subjective experiences and cartographic representations of the geo-space. Neogeographies and psychogeography both extend the perception of geographical space to the limits of the unconscious: the former in the context of the global topologies of the geo-cyber-space and the latter in the context of personal topologies. In this conceptual framework, "nomad thought" (Massumi, 1992, p.198) moves freely, in contrast to "sedentary thought", thus enriching our everyday lives with a "richesse d' instants" (Virilio, 1989, p.26). In a way, the development of neogeography is a "post-structuralist geography". In his account of such geographies, Doel (1999, p.3) quotes Fuller (1992) mentioning the "delirious cartographies" discussed by Deleuze and Guattari in their "Thousand Plateaus"; such ubiquitous cartographies may give the sense of "labyrinthic peregrination" (Doel, 1999, p.4) by rendering geography "an act, an event, a happening" in hybrid worlds (geo-and-cyber-spaces) with nomadic behaviors and practices.

Shifts from the geographic to the virtual and vice versa are reminiscent of "copresence" (Lombard & Ditton, 1997): the sensation of presence of other people in the same location, when in fact they are located in distant locations of the geo-space. This "fusion" of cyber-space and geo-space is far more important than we have hitherto realised and its nature is topological and, metaphorically perhaps, "the situationist's division of the city echoes the psychologists' "topology of the brain" (Bukatman, 2005, p.169). Furthermore, the cyber-spatial digital world converges with the geo-spatial world with AR/MR/XR in ways similar to those described by De Certeau (1984,p.93) for everyday life in the geo-space: "The everyday has a certain strangeness that does not surface, or whose surface is only its upper limit, outlining itself against the visible". And therein lies "another spatiality, an anthropological, poetic and mythic experience of space", therefore resulting in a situation where "a migrational, or metaphorical city thus slips into the clear text of the planned and readable city" (op.cit). The ways these take place is explained by De Certeau again (De Certeau, 1984, p.117) by his referring to Merlau-Ponty that there is a "geometrical space" and, simultaneously, an "anthropological space" that is a non-geometrical, non-homogeneous, anisotropic space which can be considered as the "outside" of our geometrical space (geo-space).

Where does this interplay between topology, technology, informatics and physical space lead us to? Most likely, to new spaces (physical, virtual or both) and, further still, to spaces within spaces, which means to *ever more complex topologies*. The idea by Deleuze and Guattari (1987) that there is a "striated space" (*espace strié*), metric, measured, stratified, hierarchically organized and a "smooth space" (*espace lisse*) of instability and nomadism becomes more relevant than ever before. The former applies to the geo-space and the latter to the cyber-space. Perhaps no philosophical approach to the intimate connection between the geo-space and the cyber-space has been more appropriate than this.

Globalization has brought about an unprecedented increase in the connectivity of persons, enterprises, states, goods, information, and energy flows. The word "connection" means a link between two nodes and it may be a plausible assumption that it might be one of the most important keywords of the twenty-first century. Although the concept of connectivity in geographical space may seek its origins back in Euler's "Königsberg bridges problem" (Euler, 1752), it becomes more and more obvious that some of the world's biggest problems nowadays are topological: "pipeline wars", "border disputes", "network failures" etc. translate in connections, adjacencies and links among countries, regions, geographical units, energy resources, technologies, people.

The expansion of so many networks worldwide might give the impression that increasing network size might downplay the role of topology in a network. But topology defines the very structure of information and/or telecommunication networks, regardless of how many nodes and links they have. As much as size does not matter in homeomorphisms, braids or knots, so it doesn't for the topology of networks, because topology is the template that a network adheres to, the overall plan, the inner structure that an information network faithfully observes.

And yet, despite its central role in everyday life, the topology of networks is not the only facet of the ongoing topological revolution; other concepts of topology emerge from other fields of advanced technologies that increasingly pervade everyday life. For instance, the last years have seen the growth of indoor topology that relates to problems of pathfinding and wayfinding by both humans and robots (Thrun, 1998; Lin & Lin, 2018). It is also of interest to both topology and geography that our society increasingly relies on things not located at fixed locations (Mitchell, 2003) as we have created technologies allowing "teleaction" (Manovich, 2001) and "telepresence" (Minsky, 1980). All these mean variable personal topologies.

Personal maps relate the self to the world and "boundaries are not the limits of the self, but rather they create that sense for the self" (Crang & Thrift, 2004, p.9). "Subjective cartographies" Guattari (1995, p.18,131) take place on the interface between the "actual" and the "virtual". "Event–centred singularities" (op.cit., p.7) lurk in everyday life, in consciousness and in emotions, prompting us to adopt a different cartography, an "ecosophic cartography" (op.cit., p.128) to deal with them. As connections among and with inanimate objects are anticipated through devises, sensors and big networks (big in terms of both spatial extent and data content), the decisions to connect or not (with someone or something), how and when, who with and what with, are certainly personal, but in practice they are essentially topological.

Although preferential connectivity is effectuated at the personal level, the local topological properties of the global topology of all countless connections among personal topologies might eventually become among the key determinants of the pace of progress of the humankind.

At the personal level, questions of personal topology will thus puzzle us all the more often: "With whom and what do I need to connect with?" "How will each connection change my personal topology?". And, eventually, "Do I really need a new personal topology, or am I happy with the one that I already have?"

References

Baudrillard, J. (2013). *The intelligence of evil, or the lucidity pact*. Bloomsbury.

Bonanno, G., Caldarelli, G., Lillo, F., & Mantegna, R. (2003). Topology of correlation-based minimal spanning trees in real and model markets. *Physical Review E, 68*(4), 46130.

Boss, M., Elsinger, H., Summer, M., & Thurner, S. (2004). Network topology of the interbank market. *Quantitative Finance, 4*(6), 677–684.

Bukatman, S. (2005). *Terminal identity: The virtual subject in postmodern science fiction*. Duke University Press.

Capra, F. (1996). *The web of life*. Harper Collins.

Crang, M., & Thrift, N. (2004). *Thinking space*. Routledge.

Cubitt, S. (1998). *Digital aesthetics*. Sage.

De Certeau, M. (1984). *The practice of everyday life*. University of California Press.

Debord, G. (1989). *Theory of the derive. Situationist international anthology* (p. 53). Bureau of Public Secrets.

Deleuze, G., & Guattari, F. (1987). *A thousand plateaus: Capitalism and schizophrenia*. University of Minnesota Press.

Doel, M. (1999). *Poststructuralist geographies*. Edinburgh University Press.

Elias, A. J. (2010). Psychogeography, détournement, cyberspace. *New Literary History, 41*(4), 821–845.

Euler, L. (1752 [1741]). Solutio problematis ad geometriam situs pertinentis, 1736). *Commentarii Academiae Scientiarum Imperialis Petropolitanae 8*, 128–140.

Fadoul, M. M. (2020). Modeling multi-tier heterogeneous small cell networks: Rate and coverage performance. *Telecommunication Systems, 75*, 369–382.

Fagiolo, G., Reyes, J., & Schiavo, S. (2008). On the topological properties of the World Trade Web: a weighted network analysis. *Physica A, 387*, 3868–3873.

Fagiolo, G., Reyes, J., & Schiavo, S. (2009). World-trade web: Topological properties, dynamics, and evolution. *Physical Review E, 79*(3), 36115.

Fink, T. M., & Mao, Y. (1999). Tie knots, random walks and topology. *Physica A, 276*(1–2), 109–121.

Fleischer, M. (2001). The Gaze of the Flâneur in Siegfried Kracauer's Das Ornament der Masse. *German Life and Letters, 54*(1), 10–24.

Fuller, J. (1992). Distillations-A premonitory reading of Deleuze. *Pli: Warwick Journal of Philosophy, 4*(1/2), 159–173.

Garlaschelli, D., & Loffredo, M. (2004). Fitness-dependent topological properties of the world trade web. *Physics Review Letters, 93*(18), 188701.

Garlaschelli, D., Battiston, S., Castri, M., Servedio, V., & Caldarelli, G. (2005). The scale-free topology of market investments. *Physica A, 350*, 491–499.

Gonzalez, J. A. (1995). Autotopographies. In J. R. Brahm & M. Driscoll (Eds.), *Prosthetic territories: Politics and Hypertechnologies*. Westview press.

Guattari, F. (1995). *Chaosmosis: An ethico-aesthetic paradigm*. Power Publications.

Jiang, B. (2015). The fractal nature of maps and mapping. *International Journal of Geographical Information Science, 29*(1), 159–174.

Jiang, B., & Claramunt, C. (2004). Topological analysis of urban street networks. *Environment and Planning B: Planning and Design, 31*(1), 151–162.

Jiang, B., & Liu, C. (2009). Street-based topological representations and analyses for predicting traffic flow in GIS. *International Journal of Geographic Information Science, 23*(9), 1119–1137.

Jiang, B., & Ren, Z. (2019). Geographic space as a living structure for predicting human activities using big data. *International Journal of Geographical Information Science, 33*(4), 764–779.

Jiang, B., Duan, Y., Lu, F., Yang, T., & Zhao, J. (2014). Topological structure of urban street networks from the perspective of degree correlations. *Environment and Planning B: Planning and Design, 41*(5), 813–828.

Kiyota, K. (2022). The COVID-19 pandemic and the world trade network. *Journal of Asian Economics, 78*, 101419.

Kvas, K. (2014). Ntopia: A Google street view of Psychogeography in space-times of Cybergeography. *Localities, 4*, 177–210.

Lauster, M. (2007). Walter Benjamin's myth of the Flâneur. *The Modern Language Review, 102*(1), 139–156.

Lin, W. Y., & Lin, P. H. (2018). Intelligent generation of indoor topology (i-GIT) for human indoor pathfinding based on IFC models and 3D GIS technology. *Automation in Construction, 94*, 340–359.

Lombard, M., & Ditton, T. (1997). At the heart of it all: The concept of presence. *Journal of Computer mediated Communication, 3*, 2.

Maksymyuk, T., Brych, M., & Pelishok, V. (2015). Stochastic geometry models for 5G heterogeneous mobile networks. *Smart Computing Review, 5*(2), 89–101.

Manovich, L. (2001). *The language of new media*. MIT press.

Massumi, B. (1992). *A User's guide to capitalism and schizophrenia: Deviations from Deleuze and Guattari*. MIT Press.

Mauz, C., Jaeger, M., Inkret, R., Mikac, B., & Derkacz, J. (2003). Pan-European optical transport networks: An availability-based comparison. *Photonic Network Communications, 5*(3), 203–225.

Minsky, M. (1980). Telepresence. *The Omni magazine, 2*, 45–52.

Mitchell, W. J. (2003). *Me++. The cyborg self and the Networked City*. MIT Press.

Papadimitriou, F. (2013). Mathematical modelling of land use and landscape complexity with Ultrametric topology. *Journal of Land Use Science, 8*(2), 234–254.

Pavan, C., Morais, R. M., Ferreida da Rocha, J. R., & Nolasco Pinto, A. (2010). Generating realistic optical transport network topologies. *Journal of Optical Communication Networks, 2*(1), 80–90.

Rheingold, H. (2002). *Smart mobs: The next social revolution*. Perseus.

Serrano, M. A., & Boguná, M. (2003). Topology of the world trade web. *Physical Review E, 68*(1), 015101.

Sienkiewicz, J., & Holyst, J.A. (2005). Statistical analysis of 22 public transport networks in Poland. *Arxiv: Physics*/0506074v2.

Thrun, S. (1998). Learning metric-topological maps for indoor mobile robot navigation. *Artificial Intelligence, 99*(1), 21–71.

Urry, J. (2003). *Global complexity*. Polity.

Virilio, P. (1989). *Esthétique de la Disparition*. Galilée.

Xue, Y-H., Wang, J., Li, L., He, D., & Hu, B. (2010). Optimizing transport efficiency on scale-free networks through assortative or dissortative topology. *Arxiv*: 1003.2488v1.

Correction to: Geo-Topology

Correction to:
F. Papadimitriou, *Geo-Topology*, GeoJournal Library 133, https://doi.org/10.1007/978-3-031-48185-7

The original version of this book was published with a spelling mistake in the author's last name. The original version of this book has been revised. A correction to this book can be found at https://doi.org/10.1007/978-3-031-48185-7

The updated version of this book can be found at
https://doi.org/10.1007/978-3-031-48185-7

© The Author(s), under exclusive license to Springer Nature Switzerland AG 2024

F. Papadimitriou, *Geo-Topology*, GeoJournal Library 133,
https://doi.org/10.1007/978-3-031-48185-7_15

Index

© The Editor(s) (if applicable) and The Author(s), under exclusive license to Springer Nature Switzerland AG 2023
F. Papadimitriou, *Geo-Topology*, GeoJournal Library 133,
https://doi.org/10.1007/978-3-031-48185-7

The manufacturer's authorised representative in the EU is Springer Nature Customer Service Centre GmbH, Europaplatz 3, 69115 Heidelberg, Germany. If you have any concerns regarding our products, please contact ProductSafety@springernature.com

Printed and bound by CPI Group (UK) Ltd, Croydon, CR0 4YY
01/10/2026
02225851-0002